HOW TO INVENT
A Text for Teachers and Students

HOW TO INVENT
A Text for Teachers and Students

By B. Edward Shlesinger, Jr.

IFI/PLENUM
New York • Washington, D.C. • London

Library of Congress Cataloging in Publication Data

Shlesinger, B. Edward, 1924–
    How to invent.

    Rev. ed. of: The art of successful inventing. c1973.
    Includes index.
    Summary: Presents basic principles, methods, and tools involved in the inventing
process.
    1. Inventions. [1. Inventions] I. Shlesinger, B. Edward, 1924–   . Art of successful
inventing. II. Title.
T339.S545 1987   608   87-18608
ISBN 0-306-65210-2

IFI/Plenum Data Corporation is a division of
Plenum Publishing Corporation
233 Spring Street, New York, N.Y. 10013

Printed in the United States of America

Dedicated
to
the women in my life,
my wife, daughters, and mother

PREFACE

Men and women, young and old, frequently say to me "What we need is 'a-watcha-ma-call-it', why don't you work on it?" or "I've got an idea, you ought to think about it", or "The problem with this thing is—, why can't you invent something to solve it?"

When people find that I am an inventor they somehow feel that I have the only set of keys which opens all doors. Not so! This book should provide a set for everyone. Each one of us has latent inventive ability, but most do not realize this because they imagine a few gifted people are inventive.

Only after years of inventing did I realize that there is no great mystery involved and that anyone can invent if he or she follows a checklist until all the possibilities are exhausted.

My longtime desire has been to present to teachers and students everywhere, and at every level of education, a simple, practical, and methodical approach to invention and so increase manifold the ordinary person's creative ability. This book realizes that dream.

I want to first express special thanks to my most valuable critic, my wife, Rita. Her many helpful suggestions and comments and her boundless patience over the years in no small way has made this book possible. To my law partners of many years, George Arkwright and George A. Garvey, I want to express my appreciation for their encouragement, patience, and suggestions, and I also wish to thank Gerald M. Murphy, with whom I have been long associated in the patent field, for his aid in preparing the final drawings used in the illustration in this book.

And finally, I want to thank the United States Patent and Trademark Office for technical assistance, information, and aid in processing various Patent Office documents used in this book.

And so, I hereby turn over to you this set of keys. Try them on each door you come to. You will be surprised to see how many doors this one set will open.

B. Edward Shlesinger, Jr.

TABLE OF CONTENTS

LIST OF TABLES

LIST OF PLATES

LIST OF ILLUSTRATIONS

The United States.

To all to whom these Presents shall come. Greeting.

Whereas Samuel Hopkins of the city of Philadelphia and State of Pensylvania hath discovered an Improvement, not known or used before such Discovery, in the making of Pot-ash and Pearl-ash by a new Apparatus and Process; that is to say, in the making of Pearl-ash 1st by burning the raw Ashes in a Furnace, 2d by dissolving and boiling them when so burnt in Water, 3d by drawing off and settling the Ley, and 4th by boiling the Ley into Salts which then are the true Pearl-ash, and also in the making of Pot-ash by fluxing the Pearl-ash so made as aforesaid; which Operation of burning the raw Ashes in a Furnace, preparatory to their Dissolution and boiling in Water, is new, leaves little Residuum, and produces a much greater Quantity of Salt: These are therefore in pursuance of the Act, entitled "An Act to promote the Progress of useful Arts", to grant to the said Samuel Hopkins, his Heirs, Administrators and Assigns, for the Term of fourteen Years, the sole and exclusive Right and Liberty of using, and vending to others the said Discovery, of burning the raw Ashes previous to their being dissolved and boiled in Water, according to the true Intent and Meaning of the Act aforesaid. In Testimony whereof I have caused these Letters to be made patent, and the Seal of the United States to be hereunto affixed. Given under my Hand at the City of New York this thirty first Day of July in the Year of our Lord one thousand seven hundred & Ninety.

G. Washington

City of New York July 31st 1790. –

I do hereby certify that the foregoing Letters patent were delivered to me in pursuance of the Act, entitled "An Act to promote the Progress of useful Arts"; that I have examined the same, and find them conformable to the said Act.

Edm: Randolph Attorney General for the United States.

(Endorsement on back of grant)

Delivered to the within named Samuel Hopkins this fourth day of August 1790.

Th Jefferson

First United States Patent Grant
July 31, 1790
(Reproduced from the original in the collection of the Chicago Historical Society)

1 / INTRODUCTION

Everything has a beginning;
with invention,
beginnings are everything.

The nature of invention

Invention is the very foundation of civilization. Invention is one of the strongest driving forces in human affairs and unless we have an understanding of invention, we can hardly comprehend the past, or present, or predict the future.

The rewards for inventing are many. An inventor achieves a certain amount of personal satisfaction in being first to come up with a new development. Further, when a person is known as an inventor, his position is enhanced in the community and in his place of business; and, if he has received a patent on his invention, his name becomes part of a permanent record for all posterity.

But beyond these achievements, an inventor makes possible the development of new industries, new markets and new jobs. Economically, your success can be as great as that of Chester Carlson who developed the first successful electrostatic dry copier which resulted in the formation of the Xerox Corporation, a multi-billion dollar industry, making Carlson a multi-millionaire. Carlson was not discouraged when for ten years his idea was turned down by some of the largest corporations. He persevered until a small corporation named Haloid agreed to risk capital to develop his invention. Haloid subsequently became Xerox.

Success stories in the field of invention are not rare though few may be as successful as the Carlson story. The Slinky ® toy, the Hula Hoop ®, and the Frisbee ® are typical inventions which have realized large profits for their inventors. We all have dreams of "striking it rich." Why not by inventing? We may not become millionaires and we may have only the satisfaction of being the first to create a new thing, but first steps sometimes lead to great success.

Early invention was more than likely by accident. Primitive people were very slow to accept changes and to recognize invention. In fact, early civilizations were prone to consider inventors as gods, sorcerers, or witches. Even the alchemists of the Dark and Middle Ages were looked on with apprehension. Probably the Crusades were the primary cause for spurring men on to the major inventions that resulted in the Industrial Age. Travel stimulated a desire for knowledge and vice versa, but as late as 1900, a new invention was not a common thing.

The first United States Patent was granted in 1790. By 1836 about 10,000 patents had been granted and by 1900 about 600,000. In 1970, the 3,500,000th patent issued representing millions of man-hours of time. Add this to the total number of man-hours required to produce all of the inventions of all of the other countries in the world and the sum total becomes incomprehensible. Today, few inventions are by accident. Most inventing involves a careful, deliberate, reasoning process that includes a thorough understanding of the problem, a thorough knowledge of the history involved and the materials available, and diligent experimentation. Invention begets inventions.

Although man has an inherent ability to create, he must create from something. The more facts that an individual can assimilate, the greater is his potential to invent. The average inventor receives his first Letters Patent when he or she is between 35 and 39. Corporate inventors seem to start inventing earlier than non-corporate inventors, probably because the corporate inventor's factual knowledge is highly concentrated in a certain field and pressures are greater to develop solutions to problems in their job areas. Furthermore, inventors become more prolific as they get older due primarily to two factors: 1. a continued expansion of their specific factual knowledge, and 2. an increased systemization in their thinking.

Few if any of our most prolific inventors ever stopped to analyze the methods and procedures that they used to invent.

Nov. 11, 1930. A. EINSTEIN ET AL 1,781,541
REFRIGERATION
Filed Dec. 16 1927

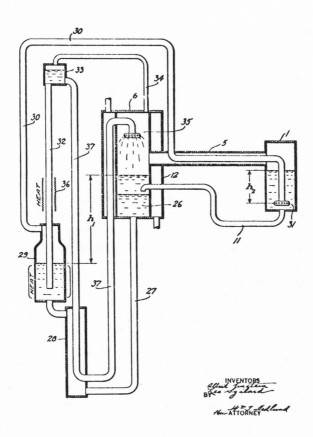

Unfortunately, such great inventors as Thomas Jefferson, Benjamin Franklin, Thomas A. Edison, Alexander Graham Bell and Robert H. Goddard never detailed the thought processes that made them great inventors so that these processes could be taught to others. Now in this Age of Invention, technology is becoming more highly sophisticated and complex requiring greater demands on time and money.

It is a well-known fact in industry that billions of dollars have been expended on research and development which have gone to naught. This is because the project inventors were unable to see the problems that would develop once the invention was made or put into production. It is not because the basic concepts were unworthy of consideration, but rather because the inventors failed to expand their thinking to include other profitable aspects of the invention. Because of a nebulous approach to invention, corporations are constantly testing and trying out the devices of inventors who have little or no system for exploring the ramifications of their basic concepts. One of the weaknesses in industry today is that few inventors look objectively at their inventions. They may be so overwhelmed by their ability to come up with a solution to a problem that they fail to look for other solutions. If they are of supervisory capacity in the corporation, the tendency is to push their concepts to the detriment of more promising concepts. In our Apollo program, for example, had it not been for aeronautical engineer John Houbolt going over the heads of planning groups to promote his Lunar-Orbit-Rendezvous-Technique, we might never have orbited the moon nor landed on it so early in the project. The problems he encountered in getting his idea accepted were basically problems of having to deal with individuals being too "sold" on their own concepts to accept someone else's more practical solution. Today inventors must take a second look and use a much broader approach to invention than was heretofore required. To be successful an inventor will need more than ingenuity. He must be willing to recognize and deal with the weaknesses and strengths of his invention, so that he may continue to pursue the best avenues of its development.

Creative man can no longer afford to wrap himself up in the mysterious cloak of the inventor. In order for man to continue to progress, he must efficiently use every possible means to increase his technical knowledge and share it with others. The growth in world population, diminution of our natural resources, and pollution are typical of the problems that face us. It is necessary that we find solutions to these and the many other problems affecting everyday life.

By following the procedures set out in this book, you can learn to invent and become a productive and efficient inventor and in so doing we shall each of us benefit, but most of all, we shall efficiently provide for the progress of our children and the generations who follow them. For teachers, this book will provide a ready check to determine whether or not all avenues available for idea development have been explored.

Using your assets to best advantage

In order to invent in an organized manner, it is not necessary that you have an elaborate setup. It is helpful to have work space in which to conduct any necessary research and experiments and assemble prototypes. You may have expensive equipment, materials, and reference works, or simply a drawing pad and pencil. Your own interests and specialities should be the factors that determine how much and what kind of equipment is necessary, if any. If your ideas require working models, and you are not equipped to produce these, you might need the services of a commercial lab or shop.

You are most likely to come up with an idea if you work in a field in which you have some familiarity. Therefore, you should work in a field suited to your interests, talents and education. It is also very important that you pick one field and stay with it rather than to try to invent in several different fields. The more information you can obtain in this one field the better are the chances for developing a successful invention.

Definition of invention

Invention is merely the correlation of at least two pieces of information. A small child immediately recognizes a comb and a triangle. A child can put these two pieces of information together and make a triangular comb. Eli Whitney put together a comb and a cylinder to invent the cotton gin.

Most invention is based on the improvement of existing conditions, developments, devices and technology. Revolutionary scientific and technological breakthroughs do occur (X-Ray; atomic energy; the transistor; the laser), but they are the exceptions to the rule.

For the purposes of this book, an invention is created when existing conditions or devices are changed.

Two kinds of invention exist: Utility and Design. A *design* invention involves a change in appearance, and not function. A pattern on a piece of silverware or carpet usually does not affect its function. A *utility* invention, however, involves a change in function, whether appearance is altered or not. When the eraser was added to the end of a wooden pencil, it afforded the pencil a new usefulness. When the golf ball was dimpled, it traveled a greater distance.

Two conditions are necessary for utility invention: 1. A difference must exist between what existed before and what was created. The difference need not be great. 2. The difference must create an advantage over the original invention; i.e. the invention at least changes the nature of the prior device or technology, in that the invention results in a gain, a new technique, a new effect, or a new physical force.

A preview of inventing—an example

In the step process set out in this book, it must be pointed out that invention can occur at any time. Once a new invention is made, it is important to continue through the remaining steps so as to fully develop all other ramifications. A review or reviews of the entire step process might bring out new ideas previously overlooked. **We should continue review-ing until we feel reasonably sure that all the possibilities have been exhausted.**

To show that you can invent, take the simple blackboard eraser.

What are its disadvantages?

1. It becomes dusty when in use for some time, getting chalk on clothes and hands.

2. It does not wipe properly when full of dust.

3. It must be cleaned by clapping against another eraser or a hard surface.

4. It does not wipe a large area.

What are some solutions?

1. To eliminate the dust we could make it a wet type of eraser. Thus we could provide the eraser with a water reser-voir.

2. To prevent its becoming full of dust we could have a changeable wiping surface. Thus we could have a reserve sup-ply of surfaces built into the eraser such as a replaceable roll of felt which can be wound from one spool to another so as to present a new wiping surface when the previously exposed wiping surface gets too dusty.

3. To simplify the cleaning process of the standard eraser, we could have a miniature vacuum cleaner installed near the blackboard. Cleaning the eraser would merely involve holding the eraser against the operating vacuum cleaner.

4. To make the eraser wipe a larger area we could double its size or provide an extensible eraser which could be used to wipe both large and small areas.

Inventing by considering the disadvantages of a device or item as set out above is but one of the many ways to invent. We will cover this approach and the many others system-atically and in detail in the following chapters.

In our study of the process of inventing, we will con-sider in detail, in the following chapters, the five major areas of study, which are:

1. Identification
2. Foundation
3. Data
4. Imagination
5. Limitations

S. LAKE.
SUBMARINE VESSEL.

No. 581,213.   Patented Apr. 20, 1897.

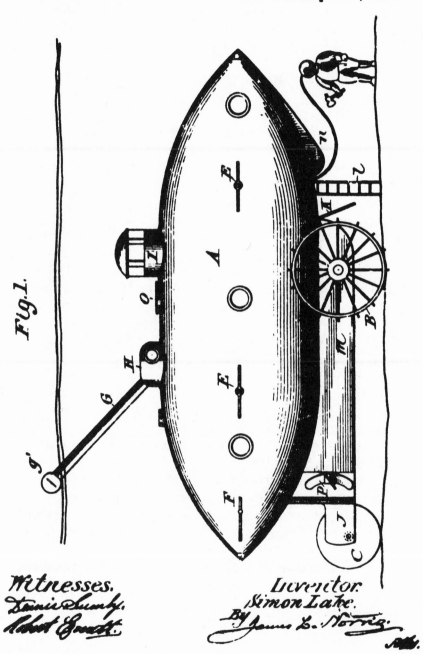

Fig.1.

2 / IDENTIFICATION – "THE PROBLEM" – HOW TO FIND A NEED AND DEVELOP A SOLUTION

"The problem"

If you wish to invent, you should first seek problems to solve. Most corporate inventors are given their problems by management who encounter problems in testing, manufacturing and marketing. For the inventor who has little familiarity with corporate problems, inventing becomes more difficult since he must ordinarily seek his problems as well as the solutions. This and succeeding chapters set out procedures for discovering problems.

The need for invention

It has been said that "Necessity is the Mother of Invention." It is immaterial whether the need be past, present or future. In the late 1800's Jules Verne wrote a novel about a trip to the moon. At the time of this novel, everyone knew that such a trip was technically impossible. But from the realistic standpoint, the book raised the basic problem—how can we get to the moon? Further research on this one question led to many problems, some intricate, some not, each requiring a so-

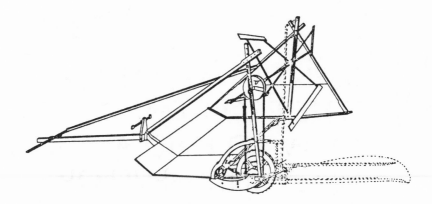

McCormick Reaper

lution. It took nearly 100 years to solve all the problems involved. Furthermore, one basic problem usually raises another
basic problem and in many cases, some solutions themselves
create problems. Thus, the problem of getting to the moon
produces the problem of returning from the moon.

Jules Verne created future needs. We can look at past or
present problems in an art and seek solutions to those which
have gone unsolved. Old problems can be solved if we look
for them. They go unsolved for many reasons; e.g. the material
necessary to construct the device was not developed at that
time. Thus a new plastic might be the solution to an old problem.

The needs that Jules Verne created could be said to be
"unnecessary." So in a sense, some inventions come from the
creation of "artificial" or "unnecessary" needs. Thus no immediate or foreseeable need may exist for an air filter system for a
city to remove impurities therefrom. We could argue that such
a device is impractical because work is now progressing on au-
tomotive and industrial anti-pollution devices which will make
such a filter system unnecessary. Nonetheless, the problem
still exists and we are free to solve it. A basic invention in this
field may come into great prominence if anti-pollution devices fail to provide adequate relief, or if the invention, even if
not itself useful, could be the stepping stone for other useful
inventions.

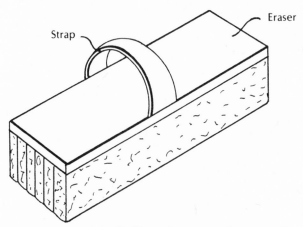

FIGURE 1.   Blackboard Eraser With Strap

Problems – complaints

One way to recognize a problem in an art is to be alert for complaints. A complaint should make us aware of a problem.

Example 1. Suppose a student constantly drops the blackboard eraser and complains that he cannot hold on to it. Obviously a problem exists. Figure 1 shows a simple solution in providing a hand strap.

Another solution would be a pressure sensitive adhesive grip.

It is therefore important for an inventor to be always on guard for complaints. For every complaint, there should be a solution.

Problem – difficult, inconvenient, and abnormal situations, recurring breakdowns – injuries

In the example above, if we did not hear a complaint, we could have recognized a problem since the student frequently dropped the eraser. You should look for the unusual. If a machine frequently breaks down, something needs improving. We can seek a solution to this problem.

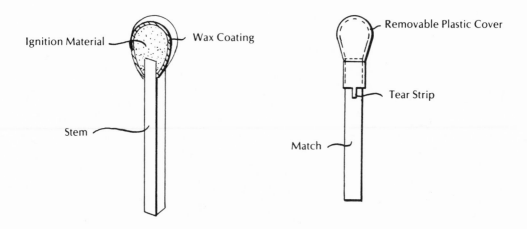

FIGURE 2.   Match Head With Wax Coating   **FIGURE 3.   Match With Removable Cover**

Example 2. A brand of book matches frequently fails to light. Analysis shows that the match head is damp from humidity in the air. Figure 2 shows the provision of a protective coating of wax for the match head.

Example 3. A similar solution, though expensive, would be to provide a removable moisture-proof plastic cover for each match head, Figure 3.

Another way to recognize a problem is to look for recurring injuries even if minor. If people occasionally or frequently trip over an item in their path, a problem exists and you should seek a solution. Relocation may be the answer, or if this is not possible or practical, then a warning device can be installed such as a flasher light or a buzzer which goes off when someone approaches the item. Safety is always important and we have a continuing need for safety inventions for machines, tools or articles. Thus if a brand of book matches occasionally ignites accidentally, further safety features may be built in to prevent serious injury.

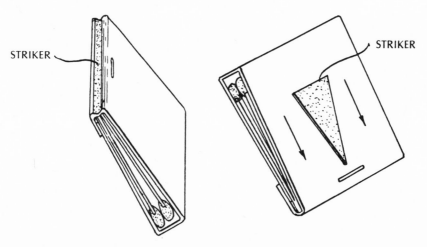

FIGURE 4. Matchbook With Recessed Striker   FIGURE 5. Matchbook With Directional Striker

Example 4. Figure 4 shows how the striker can be placed on the bottom of the matchbook to make it more difficult to accidentally ignite the book since the striker is now at the most remote place from the match heads.

Example 5. Another solution would be to have the striker designed and positioned on the back cover as shown in Figure 5. The triangular shape and the arrows tend to make most people strike from the wide part toward the narrow part and thus away from the match heads. The vertical positioning of the striker reduces the chances that flame will ignite the matches in the book. Prior striker surfaces running horizontally are more likely to have the flame of the ignited match close to the match heads with the danger of accidental ignition.

A cardinal rule in inventing is: Look at each item you see with the thought that it can be made more safe. Being safety conscious and aware of abnormal situations, recurring breakdowns and injuries will enable you to devise new inventions.

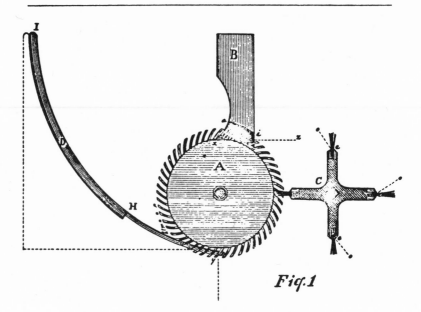

Fig.1

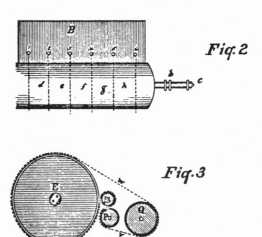

Fig.2

Fig.3

3 / FOUNDATION – HISTORY AND CLASSIFICATION AS A GUIDE TO NEW IDEAS

The initial steps of invention

The subject matter which pertains to a specific area of technology is called "art." In tracing the history and studying the classification of a certain subject you develop a feel for the art in which you wish to invent. This enables you to put that art and related arts into proper perspective. The following basic initial steps will show how development in one art affects developments in another art and thus will enable you to correlate existing technologies for the purpose of creating new inventions more readily:

1. Tracing the history of a specific area of interest.
2. Studying the classification of that area of interest and also its related areas.

Inventing through history

The history of an art tells you where we have been, where we are now and where we are going. Historical analysis can help to determine when and where a specific need may arise even if the basic invention has not been developed. Thus the need for a stabilizer was known prior to the final development of the airplane.

Table 1 shows the era in which man's most famous inventions were made. Every new invention brings with it the opportunity for its own improvement as well as the possibility

TABLE 1 – CHRONOLOGY OF SIGNIFICANT TECHNICAL ADVANCES

Era	*Shelter*	*Food*	*Clothing*	*Communication*
Primitive	Cave	Animal domestication Fishing hook	Sewing	Drum, Smoke signals
Ancient	Houses Temples	Spoilage control Sickle Plow	Weaving Spinning wheel	Horns Bells
Dark Ages and Middle Ages	Flying buttresses Stained glass windows	Windmill Irrigation		Movable type Paper making Printing press
1500 through 1800	Tile making machine Circular saw Gas lighting	Threshing machine Seed drill Iron plow	Iron needles Power loom Spinning mule Spinning jenny Cotton gin	Lead pencil Fountain pen Typewriter Stereotype Speaking tube
1801 through 1900	Portland cement Reenforced concrete Bandsaw Steel furnace Alloy steel Linoleum Incandescent lamp Gas stove Elevator	Harvester Reaper Canning Freezing Refrigerator Oleomargarine Mason jar Disc plow	Sewing machine Safety pin Shoe sewing machine Rayon Zipper Chlorine bleach	Telegraph Rotary printing press Braille printing Telephone Wireless telegraph Linotype Web printing press Quadruplex telegraph
1901 through 1925	Air conditioner Electric utility	Farm tractor Baler Silo Milking machine	Washing machine Electric washer	Radio broadcasting Radio telephone
1926 through 1940		Fluorine refrigerants	Nylon Plastic raincoats	Television Electronic computer
1941 through 1970	Prefabrication	Teflon Freeze drying	Polyester Dacron Orlon Flameproofing Permanent press fabric	Transistor Laser Xerography Communication sattelite

that this new development can be used in some other art to help in creating a basic invention or an improvement thereof. The cotton gin, a basic invention, evolved from the haircomb; whereas the chair was improved when a back was added.

Transportation	*Weapons*	*Health*	*Culture*
Foot	Throwing Piercing	Medicinal herbs	Drawings Carvings
Wheel, Cart, Roads Boat, Sail, Canals	Bronze Iron	Hospitals Sanitation	Educational 　systems Maps & Charts
Canal dredges Magnetic compass	Black powder Powder used as 　a propellant Cannon	Human anatomy 　dissection Medical societies Eye glasses	Clock Organ Harpsicord
Parachutes Canal locks Paddle-wheel boat Diving bell Steam piston 　engine Balloon	Submarine Tank Shrapnel shell	Thermometer Microscope False teeth Smallpox vaccine	Telescope Adding machine Barometer Piano
Steam locomotive Asphalt paving Steamboat Electric locomotive Motor car Vulcanized rubber Airship Gasoline Gasoline engine Diesel engine Gasoline auto- 　mobile Gas turbine Motorcycle	Torpedo Guncotton Revolver Nitroglycerine Submarine Machine gun Telescopic gun- 　sight Magazine gun Dynamite	Stethoscope X-rays Rubber dental 　plate Ophthalmoscope Cystoscope Pasturization Ether	Photography Phonograph Movie machine Player piano Kodak Disc record Gyroscope Baseball Football
Airplane Windtunnel	Military tank Depth bomb Browning gun	Typhus vaccine Vitamins Dental drill Teeth fillings	Quantum theory Relativity theory Mass spectro- 　scope
Helicopter Jet airplane	Radar Aerial bomb	Electrotherapy Artificial limbs	Uranium fission 　theory Computer
Supersonic aircraft Hovercraft Lunar landing ship	Proximity fuse ICBM Atomic bomb Hydrogen bomb	Penicillin Streptomycin Terramycin Measles, polio 　vaccines Organ transplants	Sputnik Space and under- 　water labora- 　tories

Table 2 shows the amount of activity that has taken place in regard to a given art in a particular period. Note that shelter design has intense development periods in the Ancient, Dark and Middle Ages, and from 1925 to 1970. But little change occurred in basic shelter design until the advent of the flush toilet, electrical appliances, central heating and air conditioning, each of which necessitated redesign of building structures. We will continue to see forward strides in shelter design but the rate of change in the future will probably be moderate since a modern home has few inconveniences. It is difficult to predict any great changes since we have no basic complaints. One major change can be predicted. With the depletion of oil and gas and an increase in air pollution, inventors will seek a better source of heat and light. An efficient solar furnace could be the solution. Its development could bring about major changes in our shelter design.

Probably the greatest invention strides can be predicted in the areas of Health, Transportation, Food, Clothing and Culture. We can expect great advances in Health for example, particularly in the development of new treatments for such serious illnesses as cancer and heart disease.

Table 3 takes a single item, the pocket lighter, and shows its development from earliest times. It also suggests what the future may hold for this item.

For every art in which you wish to invent, construct tables (similar to Tables 2 and 3). An historical review will assist you in coming up with new inventions. Even when you do invent a new item, its practicality and likelihood of economic success must be considered. This will be discussed in detail in another chapter.

ERA	SHELTER	FOOD	CLOTHING	COMMUNICATION	TRANSPORTATION	WEAPONS	HEALTH	CULTURE
PRIMITIVE	S	S	S	S	S	S	S	S
ANCIENT	I	M	M	S	M	M	S	M
DARK AND MIDDLE AGES	I	S	M	S	S	M	S	S
1500-1800	S	S	S	S	A	A	S	S
1800-1900	S	S	M	I	I	I	M	M
1900-1925	M	M	M	I	I	I	M	I
1925-1940	I	I	M	M	M	M	A	M
1940-1970	I	I	I	I	I	I	I	I
FUTURE 30 YEARS	M	I	I	A	M	S	I	I

LEGEND

S – SLIGHT
M – MODERATE
A – ACTIVE
I – INTENSE

TABLE 2
RATE OF HISTORICAL DEVELOPMENT OF INVENTIONS IN MAJOR CATEGORIES WITH FUTURE PREDICTIONS

ERA

ERA	
PRIMITIVE	Flint, rubbing sticks, fire, bow
ANCIENT	Flint and iron striker, magnifying glass or prism
DARK AND MIDDLE AGES	Tinder box, long fuse
1500-1700	Wooden match
1800-1900	Safety matchbox and book match
1900-1925	Liquid fuel pocket lighter
1925-1940	Platinum catalytic lighter, electric battery lighter
1940-1970	Butane gas lighter
FUTURE 30 YEARS	Reusable paper match good for many lights; miniature combination of pocket TV, radio, recorder, transcriber, watch, etc. with built in lighter; miniature fuel cell lighter; laser lighter; a lighter which will glow hot when the container is squeezed

TABLE 3
HISTORICAL DEVELOPMENT OF THE POCKET LIGHTER

Inventing through classification

Knowledge of classification will help orient your thinking with regard to related areas in a particular art or with regard to related arts thus permitting you to study all pertinent inventions with the idea of incorporating the novelty of one invention into another. Classification in its broadest sense consists of three primary classes:

1. Mechanical
2. Chemical
3. Electrical

Patents can be classified as:

1. Method or Process
2. New Use of an Old Process
3. Article
4. Machine
5. Composition
6. Design
7. Plant

Classification tables are important to an inventor and you can construct your own whenever you desire but prepared tables are readily available. One of the finest tools an inventor can use is the Manual of Classification of Patents. Most of the major countries in the world have a detailed classification system. In the United States, this manual which has several hundred major classes, is entitled "United States Manual of Classification". It is sold by the United States Government Printing Office, Washington, D. C. 20402. Plates 1 through 6 show the following pages taken from this manual:

Plate 1-Class 35-Education (page 1)

Plates 2 and 3-Class 46-Amusement Devices, Toys (pages 1 and 2)

Plate 4-Class 272-Amusement and Exercising Devices (pages 1 and 2)

Plates 5 and 6-Class 273-Amusement Devices, Games (pages 1 and 2)

Plate 1

Original Classification V. H. Todd 1934
Definitions In Bulletin 197

1	MISCELLANEOUS
2	CRYPTOGRAPHY
3	Mechanisms
4	Electrical
5	KEYBOARD
6	With display
7	PLANS AND CHARTS
8	TEACHING
9	Question and hidden answer
10	Science and mechanics
10.2	Navigation
10.4	Object detection e. g., radar
11	Vehicle operation
12	Aircraft
13	Engines and machines
14	Signaling
15	Textiles
16	Architecture
17	Surgery
18	Chemistry
19	Physics
20	Biology and taxidermy
21	Social science
22	Psychology
23	Religion and morality
24	Economics and business
25	War
26	Drawing, painting, and sculpturing
27	Designing
28	Changeable objects and pictures
28.3	Color comparison charts
28.5	Painters' mixing charts
29	Physical culture
30	Mathematics
31	Arithmetic
32	Counters
33	Abacus
34	Geometry and trigonometry
35	Language
36	Writing
37	Guides
38	For the blind
39	Reading indicating instruments
40	Geography
41	Relief maps
42	Segmental and sectional maps
43	Astronomy
44	Charts
45	Tellurions
46	Globes
47	Sidereal or celestial
48	Examination devices and methods
49	ARTICLE AND SURFACE DEMONSTRATION
50	Intercomparison
51	Interior view
52	Manufacturing stages
53	Changeable model
54	Charts
55	Material
56	Garment
57	Boot and shoe
58	With mirror
59	Toiletry
60	FURNITURE
61	ERASABLE SURFACES
62	Combined
63	Specially mounted
64	Folding
65	Frames
66	Of special materials
67	Chalk and eraser rails

69	BLOCKS AND CARDS
70	Number
71	Alphabet and word
72	Geometrical form
73	ASSEMBLY DEVICES
74	DISCS
75	SLIDES
76	MOVABLE BANDS
77	ROLLERS

Plate 2

April 1955

CLASS 46, AMUSEMENT DEVICES, TOYS

46-1

Original Classification   B. Wohlfert   1935
Subsequent Revision   C. D. Angel, N. Ansher
Definitions in Bulletin No. 204

1	MISCELLANEOUS	48	With sparking
2	TOY MONEY BOXES	49	With chromatic effects
3	Coin operated or released figure or	50	Gyroscopic
	mechanism	51	Whirlers
4	Coin depositing mechanism	52	Sounding
5	Retracting coin slot mechanism	53	Air actuated
6	SOAP BUBBLE DEVICES	54	Radiator cap ornaments
7	With reservoir	55	Figure operating
8	Power operated	56	Jet operated
9	SMOKING DEVICES	57	Convection operated
10	SPARKING	58	Pinwheels
11	CONTAINERS	59	Self-winding spindle
12	BUILDINGS	60	Aerial tops
13	Toy theatres	61	Tethered
14	HOUSE FURNISHINGS	62	Cord twist
15	Beds, chairs, tables, and cabinets	63	Sounding
16	CONSTRUCTION TOYS	64	Tops
17	Knockdown toys	65	Combined
18	Toy stores	66	With sounding means
19	Houses	67	With spinning devices
20	Log cabins	68	Spiral
21	Sheet material	69	Spring impellers
22	Figures	70	Cord impellers
23	Elements	71	Holding means
24	Blocks	72	Separable
25	Mating	73	Tips and spindles
26	With connector	74	AERIAL
27	Bars and rods	75	Autogyros and helicopters
28	Mating	76	Airplanes
29	With connector	77	Tethered
30	Sheets	78	With motor-driven propeller
31	With connector	79	Gliders
32	TOY SUPPORTS	80	With folding wings
33	TOY TELEPHONES	81	With projecting means
34	ANIMATED BOOKLET	82	Flying propellers
35	CARDS AND PICTURES	83	With cord impeller
36	Movable	84	With spring impeller
37	Changeable	85	With spiral impeller
38	CONFETTI DEVICES	86	Parachutes
39	MACHINERY	87	INFLATABLE
40	Material handling	88	Combined
41	FLUENT MATERIAL OPERATED	89	Aircraft simulations
42	Single element	90	Inflating and sealing means
43	MARBLE RUNWAYS	91	AQUATIC
44	PNEUMATIC	92	Figured
45	ELECTRIC AND MAGNETIC	93	Boats
226	with lamps	94	Submersible
227	With electric sounding means	95	Jet propelled
228	Self contained voltage source	96	Wheeled
229	Light intensity varying means	97	FIGURE WHEELED TOYS
230	Wheeled toy	98	With sounding means
231	Railway	99	Figure and wheeled support
232	Sounding means	100	Low center of gravity
233	Static induction	101	Vehicle and figure
234	Electromagnetic induction	102	Walking figure
235	Producing motion of attracted part	103	Wheeled figure
236	Permanent magnet	104	Moving figure
237	Figure head movement	105	Walking
238	Movable in spaced relation to attracted	106	Figure and wheeled support
	part	107	Moving figure
239	Interspace partition	108	Animal and rider
240	Surface for attracted part	109	Velocipede pedaling
241	Contacting armature	110	Walking
242	Suspended support	111	SOUNDING WHEELED TOYS
243	Motor operated	112	Combined
244	Vehicle reversing or steering	113	Railway rolling stock
245	Movable figure	114	Trundles and hoops
246	Radiant energy controlled	115	FIGURE TOYS
247	Self contained motor	116	Combined
46	MOLDING DEVICES	117	With sound
47	SPINNING AND WHIRLING DEVICES	118	Movable figure
		119	Movable
		120	Combined actions
		121	With dancing

SPINNING AND WHIRLING DEVICES

Plate 3

	FIGURE TOYS
	Movable
122	Along endless path
123	Animal figures
124	Birds and insects
125	Radiator cap ornaments
126	Marionette
127	Animal and rider
128	Kicking
129	Leaping and skipping
130	Acrobatic
131	Balancing
132	Climbing
133	Jumping-jack
134	Tumbling
135	Changeable feature
136	Dancing
137	Movable platform operated
138	Figure support operated
139	Link operated
140	Rotary figure support
141	Eating and nursing
142	Fighting
143	Motor operated
144	Musicians
145	Projecting
146	Jack-in-the box
147	Seesaw, swinging and rocking
148	Striking
149	Walking
150	Figure carried motor
151	Forms and dolls
152	Reptiles
153	Multiple
154	Hand
155	Self-righting
156	Of special material
157	Sheet material
158	Fabric covered
159	Jointed
160	With solid parts
161	Jointed
162	Bodies and skeletons
163	Limbs
164	Heads
165	Eyes
166	Movable eyelid
167	Movable
168	Universal
169	Horizontal axis
170	Luminous
171	Mouth
172	Wigs
173	Joints
174	SOUNDING TOYS
175	Combined
176	Detonating
177	Multiple sounders
178	Air actuated sounder
179	Whistles and sirens
180	Vibratory reed
181	Horns
182	Voice actuated diaphragm
183	Voices
184	Valved
185	Rotary
186	Pivoted
187	Reciprocating
188	Elements
189	Vibrators
190	Tick-tacks
191	Percussion
192	Ratchet actuated
193	Rattles
194	Film bursters
195	Self-feeding

	SOUNDING TOYS
196	Detonating
197	Canes
198	Magazine type
199	Projecting
200	Projectiles
201	WHEELED TOYS
202	Combined
204	With wheel actuated devices
205	Trundles
206	With driving means
207	Eccentric weight
208	With motor lock
209	Inertia
210	Remote control
211	Erratic movement
212	Reversing
213	Self steering
214	Dumping
215	Fire apparatus
216	Railways
217	Locomotives
218	Cars
219	Tractors and tanks
220	Trundles and hoops
221	Elements
222	Bodies
223	Demountable

Plate 4

July 1961

CLASS 272, AMUSEMENT AND EXERCISING DEVICES

Original Classification   M. L. Whitney   1924
Subsequent Revision N. T. Ball, E. C. Darsch,
J. W. Love, J. C. Mitchell, V. A. Morrison
Definitions in Bulletin No. 345

1	AMUSEMENT
2	Houses
3	Arenas
4	Racing
5	Horse
6	Elevating devices
7	Combined roundabouts
8	Illusions
8.5	By transparent reflector
9	Stage
10	Special projected picture or light effects
11	Settings
12	Rapid movement
13	Mirror
14	Sound imitation
15	Rain, snow and fire
16	Trip simulation
17	In passenger-carrying devices
18	With projected picture scenery
19	Maze or labyrinth
20	Pyrotechnic display
21	Stage appliances
22	Shifting scenery
23	Guides, braces and clips
24	Aerial suspension devices
25	Properties
26	Stage tanks
27	Initiating devices
28	Roundabouts
29	Combined with transporting vehicle
30	Combined seesaw
31	Toy
32	Marine
33	Occupant propelled
34	Auto-propelled carriages
35	Free carriage
36	Vertical and horizontal axes
37	Plural vertical axes
38	Plural horizontal axes
39	Vertical axis only
40	Suspended vehicle or rider support
41	Circular swings
42	With rotating platform
43	Rotating vehicle or rider support and stationary track or platform
44	Vertically undulating track or platform
45	Horizontally undulating track or platform
46	Rotating disk, ring or bowl
47	Concentric rings or disks
48	With vehicle or rider supports
49	Horizontal axis only
50	Gyrating axis
51	Inclined axis
52	Hobby horses
52.5	Combined or convertible
53.1	With actuating means
53.2	With movable elements
54	Seesaws
55	One person
56	Rocking support
56.5	Slides
85	Swings
86	Motor operated
87	Hand and foot operator

	AMUSEMENT
	Swings
88	Hand operator
89	Horizontally reciprocating
90	Cable grasp
91	Pulley mounted
92	Foot operator with separate suspender
57	EXERCISING
58	Chairs and sofas
59	Field sports
60	Gymnastic
61	Trapezes and rings
62	Horizontal bars
63	Parallel bars
64	Vaulting horses
65	Projectors
66	Spring boards
67	Hand and wrist
68	Grips
69	Tread mill
70	Walking or skating
70.1	Stilts
70.2	Steps
70.3	Occupant propelled frame
70.4	Armpit engaging
71	Swimming
72	Rowing
73	Bicycle
74	Skipping
75	Ropes
76	Striking
77	Striking bags
78	Supports
79	Push and pull
80	User attached
81	Weights
82	Springs
83	Metallic
84	Dumbbells and clubs

Plate 5

273-1   CLASS 273, AMUSEMENT DEVICES, GAMES   April 1960

Orig. Class. M. L. Whitney 1924
Sub. Rev. C. D. Angel, N. Ansher,
    E. L. Bell, E. C. Blunk
    V. H. Sweeney
Definitions in Bulletin No. 346

1	MISCELLANEOUS
1.5	BASKETBALL
2	BILLIARDS AND POOL
3	Tables
4	Convertible billiard and pool
5	Convertible furniture
6	Beds
7	Cloth securing devices
8	Rails
9	Cushions
10	Table attached ball racks
11	With pocket connected conveying chutes
12	Pockets
13	Dust covers
14	Special game attachments and accessories
17	Cue chalkers
18	Holders
19	Automatic
20	Tethered
21	Fixed
22	Ball spotting racks
23	Cue rests
24	Hand attached
25	BASEBALL
26	Practice devices
27	Field covers
28	Ball curvers
29	TENNIS
30	Table
31	Court marking strips
32	GOLF
176	Condensed, miniature or restricted courses
34	Putting holes and hole markers
177	Simulated receivers
178	Holes and depressions
179	With ball return
180	Plural pockets
181	Targets
182	With ball return
35	Practice devices
183	With indicator, alarm, recorder or register
184	Struck-projectile responsive
185	Projectile direction
186	Club-swing responsive
187	Stance
188	Body guides or restraints
189	Arm
190	Head
191	Club guides
192	Putting
193	Club
194	Modified standard type or accessories
195	Mats and tees
196	Tethered projectile
197	Rotatable
198	By flexible filament
199	Projectile
200	Tethered
33	Tees
201	With ball feeding means
202	Adjustable height
203	Plural interchangeable or rearrangeable

	GOLF
	Tees
204	Horizontally swingable
205	Of ribbon-like sheet material
206	Of wire
207	Tilt top
208	Tethered
209	Restrained by weight
210	Foldable blanks or assembled from blanks
211	Plural points in ball engaging surface
212	Diverse material head and shank
37	BOWLING
38	Mechanically projected ball
39	Changed or curved ball path
40	Tethered bowling ball
41	Special targets
42	Mechanical pin setters
43	With mechanical pin gatherers
44	Tethered pin
45	Tethered to bed
46	Pin spotters
47	Ball return
48	Automatic
49	Mechanically operated
50	Foul indicators
51	Beds and gutters
52	Pin spots
53	Back stops
54	Accessories
55	FOOTBALL
56	CROQUET
57	Tables and boards
58	BALLS
59	Billiard and pool
60	Baseball
61	Tennis
62	Golf
213	With alarm, location or indicator means
214	Center expanded or under compression
215	Injected expanded fluent material
216	Tensioned elastic windings
217	Mechanical bond between encompassing points
218	Unitary structure
219	Buoyant
220	Center
221	With air pockets
222	Wound
223	Spider or skeleton core
224	Exclusively
225	With diverse type layer
226	Plural windings
227	Rubber
228	Plural layer
229	Fibrous
230	Core
231	Liquid
232	Surface configurations
233	Cover
234	Reinforced
235	Materials
63	Bowling
64	Hand grips
65	Foot ball and basket ball
67	BATS, MALLETS RACQUETES, CUES, PINS AND BILLIES
68	Billiard cues
69	Mechanically operated

	BATS, MALLETS, RACQUETES, CUES, PINS AND BILLIES
	Billiard cues
70	Tips and tip fasteners
71	Clamps and presses
72	Baseball
73	Tennis
74	Presses and cases
75	Handle grips
76	Table
77	Golf
162	Combined
163	Ball and club alignment means
164	Integral with club
81	Handles and hand grips
165	Removable grip positioner
166	Player worn gripper
81.2	Quick detachable or longitudinally adjustable
81.3	Angularly related handle and club shaft axes
81.4	Finger-conforming configurations
81.5	Spiral windings
81.6	With spiral grooves and/or elevations
79	Adjustable ball engaging face
80	Shafts and shaft fastenings
80.1	Quick detachable or angularly adjustable head and shaft
80.2	Head and shaft joints
80.3	With common covering
80.4	Cushioned interspace
80.5	Shaft surrounding and hosel end abutting sleeve
80.6	Deformed shaft wall
80.7	Threaded shaft
80.8	With core or plug
80.9	Longitudinally seamed or sectioned shafts
167	Heads
168	Plural ball contacting faces
78	Resilient faces
169	With particular weight means
170	Shifting weight during swing
171	Adjustable by addition, subtraction or interchange
172	By modification of sole plates
173	Plates and inserts
174	Ground engaging
175	Non-planar ball contacting faces
82	Bowling pins
83	Croquet
84	Police clubs
85	SIMULATED GAMES
86	Racing or time determinative contests
87	Golf
87.2	Projector
87.4	Manikin type
88	Baseball
89	Mechanically pitched and batted ball

Plate 6

	SIMULATED GAMES	148	GAME ACCESSORIES
	Baseball	149	Card shufflers and dealers
90	Manually batted ball	150	Hand holders
91	Moving surface	151	Duplicate games
93	With chance element	152	Playing cards
94	Football	152.1	Structure
95	AERIAL PROJECTILE	152.2	Back indicia
96	Combined projector and	152.3	Transaction represen-
	catcher		tations
97	Tethered projectile	152.31	Game
98	Tethered projectile and	152.4	Suits
	target	152.41	Composite indicia
99	Tethered ring	152.42	Scoring
100	Peg and ring	152.43	Indexing
101	Combined target and pro-	152.44	Accessory cards
	jector	152.5	Rearranged basic indicia
101.1	Light ray projector	152.6	Political and/or geographic
101.2	Simulated projector	152.7	Equation, word or sen-
102	Targets		tence forming
102.1	Combined with additional	153	PUZZLES
	indicator	154	Balancing ovoids
102.2	Electrical	155	Folding and relatively
102.3	Prize dispensing		movable strips and
102.4	With projectile stop		disks
103	Projectile return	156	Take-aparts and put-
104	Pegged		togethers
105	Pocketed	157	Geometrical figures,
105.1	Projected picture		pictures, and maps
105.2	Moving	158	Bent wire
105.3	Airplane towed	159	Flexible cord or strip
105.4	Flying	160	Mortised blocks
105.5	Ball	161	FORTUNE-TELLING DE-
105.6	Handling and manipulation		VICES
106	Projectiles		
106.5	Arrows and darts		
108	SURFACE PROJECTILE		
109	Moving surface		
110	Pivoted		
111	Pivoted gate		
112	Spiral surface		
113	Pocketed		
114	Mercury globule		
115	Surface pockets		
116	Hazard pockets		
117	Moving pockets		
118	Ball games		
119	Combined with projector		
120	Gravity projectors		
121	Return course		
122	Ball return		
123	Pocketed		
124	Return course		
125	Ball return		
126	Disk or ring games		
127	Targets		
128	Projectiles		
129	Projectors		
130	WITH GAME BOARDS		
131	Definitely movable game		
	pieces		
132	Shifting arrangements		
133	Jumping only		
134	Chance controlled		
135	Chance controlled		
136	Game boards and tables		
137	Game pieces		
138	CHANCE DEVICES		
139	Chance selection		
140	Fish ponds		
141	Rotating pointer		
142	Rotating disk		
143	Edge indication		
144	Lot mixers and dispensers		
145	Dice agitators		
146	Dice		
147	Tops		

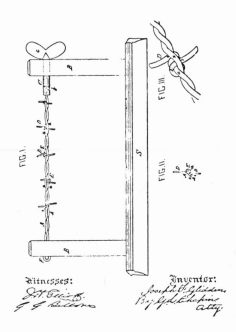

J. F. GLIDDEN.
Wire-Fences.

No.157,124.

Patented Nov. 24, 1874.

Witnesses:

Inventor:

The various arts are broken into specific subclasses and in-dented subclasses. Referring to Class 35, Education, (Plate 1), we note that subclass 8 is Teaching. A specific subclass of Teaching would be Mathematics, indented subclass 30. Sub-class 30 is further divided into additional subclasses 31 through 34. Thus ABACUS in subclass 33 can be found under Mathe-matics, Arithmetic Counters, under Teaching, Subclass 8 of Class 35, Education.

8	Teaching
30	Mathematics
31	Arithmetic
32	Counter
33	Abacus

Taking a specific example, suppose that our field of in-vention relates to Teaching, Physical Culture, Class 35, Subclass 29 (Plate 1). By reviewing all the subclasses in Class 35, you will see that various methods of teaching are immediately obvious, e.g.,

1. Keyboard Devices with Display—Subclass 6 or

2. Question and Hidden Answer Display—Subclass 9 and so on.

You can generate ideas for inventing by studying, inter-relating and correlating related classifications. Thus you might study Class 46 (Plates 2 and 3) as it might pertain to Physical Culture, Class 35, Subclass 29 (Plate 1) with the object of add-ing new ideas, concepts and devices to your background knowledge of Physical Culture. Briefly running down the sub-classes in Class 46 (Plates 2 and 3) brings novel ideas quickly to mind. Subclass 34, Animated Booklet, for example gives you a means for teaching Physical Culture. Other subclasses and their indented subclasses give you additional aspects of teach-ing such as by movable and sounding toys. Considering Class 272 (Plate 4), you now see the broad field of exercising from Chairs, Subclass 58 to Clubs, Subclass 84. Note also the related aspects of Amusement Devices from Subclass 1 to Subclass 92. Similarly, Class 273 (Plates 5 and 6) covers in detail many as-pects of the equipment used in Physical Culture.

With the Manual of Classification, you can quickly and easily obtain systematized knowledge on any one subject. Ideas gleaned from a review of classification can be mentally tested on your basic problem to see if they look promising. If they do, further study and development will permit you to se-lect those most applicable. *It is therefor important that you classify your invention area and relate it to other areas of de-velopment. You may want to create a classification system of your own particularly when others may be inadequate.* In classifying and studying classification, you will be made aware of new and different approaches to invention.

No. 821,393.

PATENTED MAY 22, 1906.

O. & W. WRIGHT.
FLYING MACHINE.
APPLICATION FILED MAR. 23, 1903.

3 SHEETS—SHEET 1.

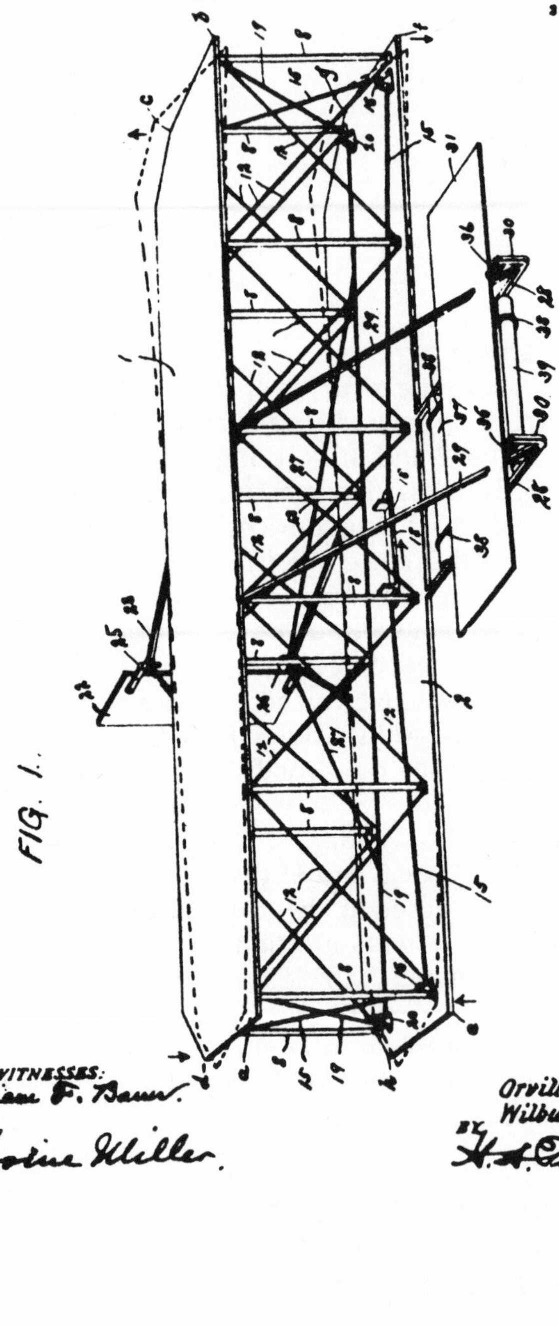

FIG. 1.

INVENTORS.
Orville Wright
Wilbur Wright
BY
H. A. Toulmin
ATTORNEY.

4 / DATA—
WHAT IS IT?

Asking the right questions

If you ask the right questions, you will more likely obtain the required information needed to invent. Without adequate information on a certain subject, you will possibly overlook important aspects necessary to achieve superior results. General questions are not sufficient. You must ask many detailed and specific questions about a problem area in which you wish to invent. You must ask these primary questions:
1. What is it?
2. Why does it exist?
3. How and why does it work?
4. When, where and how is it used?
5. How is it described?

In this chapter we will explore in detail the question "What is it?"

Qualitative and quantitative analysis

As inventor should know all facets of his subject. Analysis is of utmost importance. You should study a subject in detail as to quality and quantity. Quality in general involves "kind." You should examine the types of materials or members used

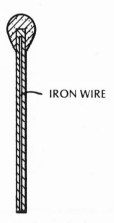

IRON WIRE

FIGURE 6.   Match With Wire to Retard Burning

with the object of finding an improvement by using a different substance, shape, or size or the like. Quantity in general involves "amount." You should study carefully the amount of a substance or the number of items used and consider whether a change in amount by adding or subtracting, increasing or decreasing will improve the subject under consideration. Since most inventions are improvements on existing inventions, your analysis should consider the following both from the quantitative as well as the qualitative standpoint:

1.   Composition
2.   Appearance (size, shape and color)
3.   Weight
4.   Other qualitites and characeristics

When we consider each of these aspects in detail, we have begun to answer the question "What is it?" Having determined what it is, you can then ask *"Can we change these characteristics in a way so as to produce a new and better result?"*

Composition

Analysis of composition can be very detailed. In the case of the matchbook, you may halt your analysis at major components but you could make a more detailed analysis such as determining the exact chemical composition of each part. The major physical components of the matchbook are the cover, the striker, the match and the fastener. The cover and most of the match are made from paper, the fastener from metal, and the striker from a chemical composition. The match also includes the igniter material.

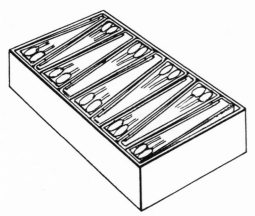

FIGURE 7. **Package of Matchbooks**

Example 6. You can consider changing the composition of any one or all of the various components of the matchbook. Instead of a metal fastener, you could use glue. You could waterproof the striker and match head by changing the ignition material. Match stems are made from paper. Could they or portions thereof be made from metal? Your first answer might be no, but let us apply this concept. You could use certain metal powders incorporated in the paper to increase the heat of the flame, make it burn brighter, slower, faster or hotter. Powdered magnesium would increase the heat and make the flame brighter. A thin wire of steel or iron incorporated into each paper match, Figure 6, would slow down the rate of burn.

Appearance

A change of appearance frequently results in a new invention. You should consider the size, shape and color of each item not only as you see it under normal conditions but also as you see it under other conditions such as radiation, magnification, stress and light variation. The drinking glass, thimble and wastebasket have the same shape, but a different function.

Example 7. A book of matches has a triangular shape because the match heads bulge out from the sides of the match stem. Packaging requires inversion of every other matchbook as in Figure 7.

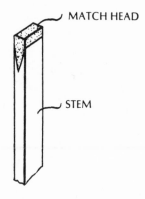

FIGURE 8.   Flat Match Head

Such an inversion requires special packaging equipment and added expense. If we make the matchbook rectangular we do not need to invert, but to make the matchbook rectangular, we must do away with the bulge of the match heads so that the match heads do not extend out from the walls of the match stem.

Example 8. Figure 8 shows a new style of match head.

Note that the paper extends outwardly on both sides of the ignition material whereas in the past, the ignition material surrounded the paper. Because of this arrangement, the flame burns outwardly into the paper and not inwardly assuring a more positive light. This single improvement solves the problem of packaging and also makes the match easier to light.

The color of an article may produce a new effect. Thus, for many years it has been known that silver light reflectors could be used to increase plant growth by concentrating the light on a plant. It was recently found that different colored reflectors had different effects on plant growth and that yellow reflectors actually stunted the growth of plants whereas red reflectors actually increased the growth of plants over that of silver

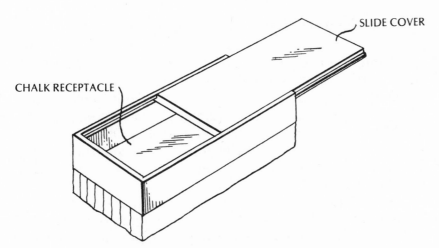

FIGURE 9.   Eraser With Covered Receptacle

reflectors. Black items tend to absorb more heat than white items. Some colors reflect better than other colors and some are more eye catching than others. Invention may result from a mere change in color.

Weight

Weight is an important aspect of invention. If an item has a specific weight or density, a change could lead to a new invention. An item might be quite light and have a limited use. By increasing the weight of the item through change of materials or adding additional material, you can obtain a dual function. Increasing the weight of a vehicle tends to stabilize the vehicle as long as the weight gives a low center of gravity. A high center of gravity makes the vehicle much more unstable. Experimentation may be required to provide the proper stability. Thousands of inventions have come about merely by a change in the weight of an object. Making certain parts of an object hollow reduces the weight and permits greater portability. Weight is an important factor as far as flight or boating is concerned. Reduction in weight may produce a major invention in an industry such as the aircraft industry.

Example 9. Figure 9 shows a blackboard eraser in which the weight of the eraser has been reduced by hollowing the upper handgrip portion of the eraser. A sliding cover provides a receptacle for holding chalk.

Other qualities and characteristics

There are many other analytical qualities and characteristics which you should consider when asking "What is it?" Some items are radioactive. The amount of radiation they give off may be an important consideration because it may require shielding or the like for protection purposes. Other items may by their nature exist in the solid, liquid or gaseous state only at certain temperatures or pressures. Changing these specific qualities may result in a new invention.

In general, knowledge of the physical and quantitative qualities of an art is important not only in understanding what it is, but also in understanding more deeply the other basic questions which must be asked and which will be covered in subsequent chapters in detail.

5 / DATA—
WHY DOES IT EXIST?

"The Tree System"

Classification discussed in Chapter 3, may answer many questions including certain aspects of the question "Why does it exist?" In order to further assist you in understanding existence, you can use "The Tree System." In this system, you make a diagram in which the object of interest becomes the trunk of the tree. The next step is to explore the historical background of the object. Your historical findings become the roots, and by considering and projecting the basic principles, you can develop new concepts which become the limbs and branches.

Example 10. Figure 10 shows The Tree System development for the blackboard:

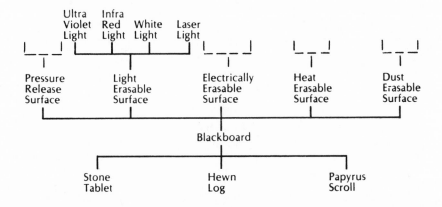

FIGURE 10.   The Tree System

Note that the historical roots of the blackboard are the stone tablet, the hewn log and the papyrus scroll. The inscriptions on the early developments were not easily erased. The blackboard is superior to these earlier inventions because it is easily erased.

If you ask what new surfaces and system can be used which are easily erased, your imagination should suggest further developments in this field. These developments are the limbs of the tree. Each limb can be examined further to produce additional branches as indicated by the dash lines. Thus the light-extinguishing surface might be the limb for such branch inventions as those that would use ultra violet (black), infra red, white and laser light; phosphorescent and a fluorescent surface materials; and opaque, translucent and transparent surfaces.

By using "The Tree System," you can develop new thoughts and ideas. If you can establish the roots, you can then consider the broad aspects of other developments leading from the basic invention. Once you understand why an invention exists, have a thorough understanding and develop basic principles, you can more readily anticipate and create improvements in that art.

Primary objects of invention

An invention exists because an inventor had certain objects in mind. Generally an invention provides a number of advantages over its predecessor. However, Rube Goldberg, the cartoonist, created many novel ideas, but his object was to produce humor by making a simple device operate in a complex way. His inventions were not practical. Practicality is therefore an important consideration in inventing, but you should not overlook the fact that such aspects as appearance may sell items even if the design may be impractical. Thus not every chair is bought for comfort alone. All chairs are functional, but decor may influence a person to purchase an attractive chair rather than a comfortable chair. You should endeavour to build attractiveness as well as practicality into an item to create advantages over prior developments.

The primary objectives of most inventions are to correct disadvantages in, and produce advantages over, the prior art.

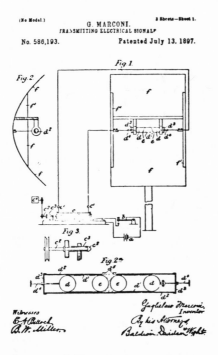

A simple procedure to follow in inventing is to enumerate all the advantages and disadvantages of an existing invention on which you are working to improve. Not only should you consider the present existing art but you should also consider the earlier developments. After listing the advantages and disadvantages you should ask yourself two questions:

1. Can I improve the advantages?
2. Can I correct the disadvantages?

The matchbook has the following major advantages:

1. Light weight
2. Compact
3. Contains a plurality of matches
4. Used for advertising

The matchbook has the following major disadvantages:

1. Accidentally ignited
2. The matches create a disposal problem

Advantages

Advantages 1 and 2. You can reduce the weight of the matchbook by using lighter material or removing some material, thus reducing costs.

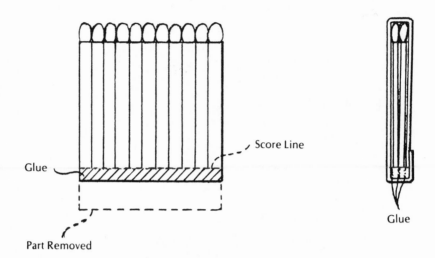

Score Line

Glue

Part Removed

Glue

FIGURES 11 and 12. Reduced Weight Matchbook

Example 11. Figures 11 and 12 show reduction in weight by removing most of the match tab base and cementing the remaining tabs to each other and to the cover. Providing a score line at the base of the matches permits easy removal of the individual matches.

Reduction in weight can cut costs by decreasing the amount of material required. The glue eliminates the need for staples. Note the package is shorter and more compact.

Advantage 3. You can increase the number of matches by increasing the size of the book.

Example 12. One way of increasing the size of the book and doubling the number of matches is shown in Figure 13.

Use of the five senses

Advantage 4. One way to improve an invention is to consider application to any one or all of the five senses:

1. Sight
2. Smell
3. Touch
4. Taste
5. Hearing

FIGURE 13.   Double Book Match

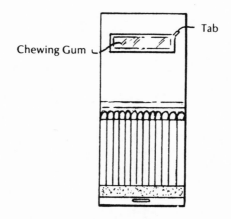

FIGURE 14.   Matchbook Containing Chewing Gum

The matchbook usually relies on visual advertising. However, you could advertise perfumes by scenting the cover (smell) or market textiles by having a fabric applied to the cover such as velvet so as to appeal to the sense of touch.

Example 13. A particular flavored chewing gum could be promoted by incorporating a wrapped piece in each matchbook as in Figure 14 (taste).

Example 14. A matchbook can be adapted to use the sound medium for advertising. Figure 15 shows a matchbook incorporating a recording surface on the inside cover. Running a pointed object over the grooved surface creates a sound such as a voice advertisement using the same principle as a phonograph record.

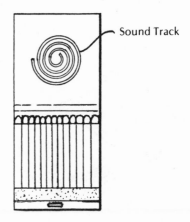

FIGURE 15. Matchbook With Recording

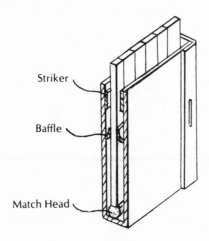

**FIGURE 16. Matchbook With Cut-Away of Accidental Ignition
Safety Feature**

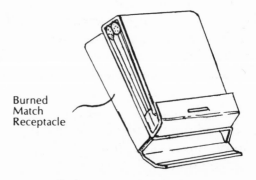

FIGURE 17. Matchbook With Burned Match Receptacle

Disadvantages

Disadvantage 1. The remaining matches in a matchbook frequently ignite accidentally.

Example 15. To prevent the matchbook from accidentally igniting when striking a match, you can have the matches held between the covers so that they will be ignited only when drawn out. See Figure 16.

Note that you ignite the matches by withdrawing them from the book thus engaging the striker surfaces on the inside of the book. The baffles prevent the flame from backing down and igniting the other matches if the matches are withdrawn slowly. The baffle material can be sponge-like to expand into the cavity left by the removed match.

Disadvantage 2. The ordinary matchbook has no built in match disposal system.

Example 16. You can correct the problem of match disposal in the manner shown in Figure 17.

Note that the receptacle has a tuck-in flap to keep the used matches from spilling out. The receptacle can also be used for cigar or cigarette ashes.

You should make your lists of advantages and disadvantages in your area of interest as detailed as possible. Clear knowledge of the objectives of the prior art and the reasons for its existence makes it easier for us to think of ways of improving it.

T. A. EDISON.
Phonograph or Speaking Machine.

No. 200,521.   Patented Feb. 19, 1878.

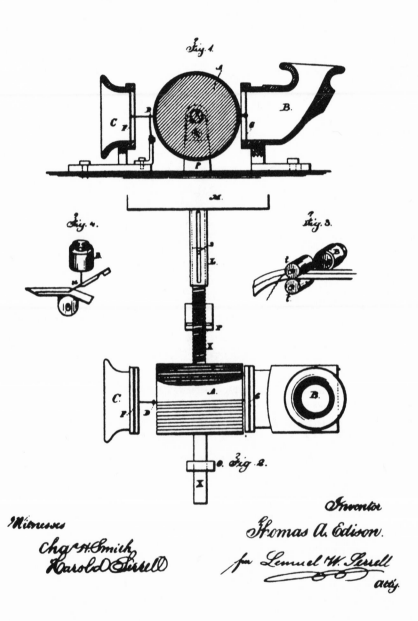

Fig. 1.

Fig. 4.   *Fig. 3.*

Fig. 2.

Witnesses   Inventor
Chas. H. Smith   Thomas A. Edison
Harold Serrell   per Lemuel W. Serrell
                                        att'y.

6 / DATA—FUNCTIONS, PRINCIPLES AND USE

How and why does it work?

To improve the prior art you should have an understanding of the related operating functions and principles. You should never proceed under assumptions. You should know in detail how each individual development operates and the principles of operation. Many people know how to use a device but fail to understand how and why it operates.

Example 17. In the most common form of the basic lever system, Figure 18, the force applied to the lever at point P pivots the lever about the fulcrum F thus moving the weight W. The illustration appears simple enough but consider the following factors:

1. The length of the lever
2. The height of the fulcrum
3. The relative position of the weight and fulcrum
4. The strength of the lever
5. The magnitude of the applied force

If the lever is not sturdy enough it will break. If the fulcrum is positioned incorrectly it will be difficult, if not impossible, to manually shift the weight. When these questions have been answered, you will be able to produce the most efficient lever system.

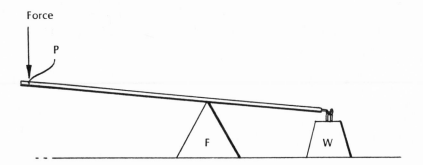

FIGURE 18. Lever System

You can understand how something works but in addition it is helpful to know why it works. The principles or rules which govern its operation should also be studied for insight into what must be done to improve it.

Example 18. You know that a match lights when you strike it against a striker with a certain pressure. The principle is that the friction created between the match and the striker produces sufficient heat between them to ignite the match. With the safety match, a special surface is also necessary so that the combination of ingredients of the striker and the match will react to cause ignition. Friction pressure is the principle involved. Can other types of pressure produce a similar result? Thus, certain chemicals explode if a shock wave is sent through them. This knowledge might enable you to create a match that would ignite through vibration. Thus we could use a tuning fork to light such matches. See Figure 19.

When placed between the tines of the vibrating fork, the match will ignite.

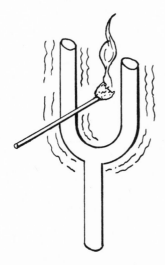

**FIGURE 19. Tuning Fork Igniting a Match
by Sonic Wave Vibration**

When is it used?

If you are to improve an art, you must frequently relate it to a definite period of time. Many inventions exist solely for the purpose of operating at or during a specific time. Thus many inventions are designed principally for operation at a specific time such as every hour, night, day, month, and so on. You should ask "When is it used?" and "When can't it be used?" A later chapter will discuss inventions dealing with specific durations of time.

Example 19. The ordinary conditions for use of the blackboard require the presence of natural or artificial light. Thus a blackboard is generally useless in the dark. What if you desired to work without ordinary room lighting? One way to solve this problem would be to use chalk which will glow in the dark. The blackboard borders could be covered with luminescent paint or plastic so as to outline the board in the darkness.

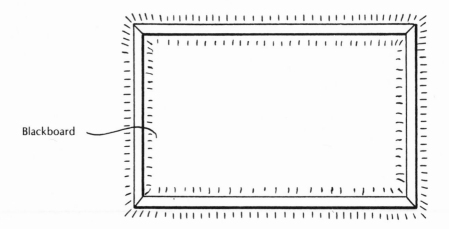

Blackboard

FIGURE 20.   Blackboard With Luminescent Border and Writing

Example 20. Sometimes a time function can be incorporated into an item. An alarm clock is typical of an invention having such a built-in function. Suppose you desire to have the blackboard cleaned every hour so as to provide a clean slate for the incoming teacher. Figure 21 shows how this can be done with an automatic eraser which will clean the board each hour.

The eraser is mounted on tracks at the top and bottom of the writing surface. A reversible motor with timer drives the eraser from one end of the board to the other. A foot control permits the instructor to shift the eraser left or right when desired.

It is always good to consider the past, present and future when asking "When is it used?" Occasionally developments were used in a different manner in the past. Marbles were first used as toys but later were used also as ball bearings. Past and present use may give some clue leading to future uses and developments. The period of use is therefore an important factor in inventing.

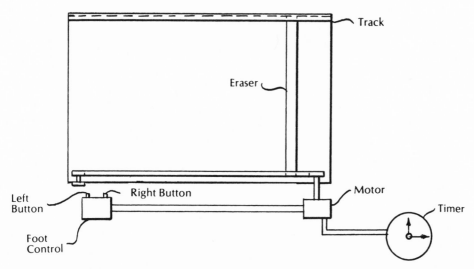

FIGURE 21. Blackboard Eraser With Timer and Foot Control

Where is it used?

An invention is frequently limited by the place in which it is used. Location may be critical. The question "Where is it used?" poses the question "Where can't it be used?" Answers to these questions will enable you to explore the possibilities of making it usable in previously unusable areas.

Geographical areas may have a bearing on specific use. Climate may play an important factor. Some items useable in one country may not be useable in another without incorporating something into or changing the environment to account for geographic as well as climatic changes. Remember that gravity, moisture, dirt temperature, pressure, radiation, magnetic forces, wind, light, dust, and the like may have a bearing on the operability of an invention.

Example 21. Blackboard chalk may not be used under water. Incorporating a white wax base into the chalk would permit such use since it would bind the chalk against dissolution and make it adhere to the blackboard. Similarly, matches may be difficult to light in windy places. The problem of correcting this might lead you to consider a wind guard for matches. See Figure 22.

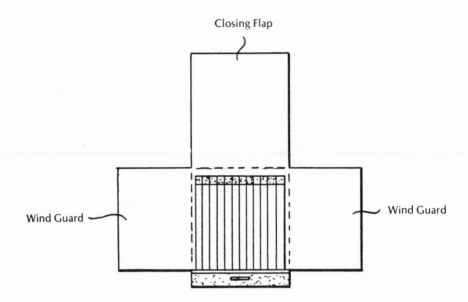

FIGURE 22. Matchbook With Wind Guard

The side flaps serve as wind guards and fold in over the matches when not in use.

How is it used?

An inventor should study all the various ways in which an invention is used in order to improve the invention and apply the concept to other fields. A thorough understanding of the method of operation might lead to a fundamental improvement in operation.

Example 22. A blackboard eraser is moved by hand in a sweeping arc-like stroke when used on the board. The stroke is similar to that of a windshield wiper. You can apply this action to a mechanism to which the eraser is attached for the purpose of cleaning the board as shown in Figure 23.

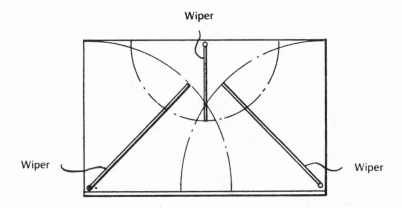

FIGURE 23. Blackboard With Wiper Blades

Note the three blades will cover the entire board. All blades could be driven from a single motor and timed so that they would not interfere with each other. When not in use they would lie out of the way against the side and upper edges of the board. Thus you have converted a manual system to a machine-operated system.

Example 23. Another illustration is the study of the action a person takes using a blackboard. One writes at the highest point down and from left to right walking back and forth shifting the arm up and down while writing. Suppose you desire to eliminate the walking and reaching to enable a person to stand in one position at all times without unnecessary stretching. Figure 24 shows how this can be done.

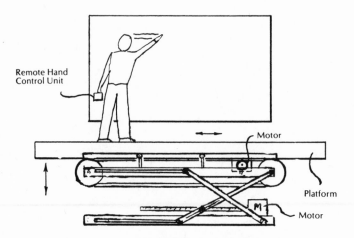

FIGURE 24.   Shifting Platform for Blackboard

The platform is raised and lowered by means of the motorized jack. The platform is also shifted right or left by the motor. The teacher can operate the motors to shift the platform up or down, right or left by a remote control hand unit. With such a device, a teacher can traverse the entire board without moving from his spot on the platform.

Who will be the user?

Many inventions are designed for a specific person or group of persons. Consider such characteristics as age, height, weight, sex, color, agility and other physical features when solving problems involving people.

Example 24. The drinking vessel shown in Figure 25 illustrates how recognizing a specific need in toddler's drinking created a cup with a spout to sip on.

In summary, you should be thoroughly conversant with each new development project knowing in detail the **hows, whys, whens** *and* **whos** *as they relate to* **functions, principles** *and* **use.**

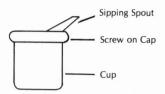

FIGURE 25.   Toddler's Drinking Cup

7 / IMAGINATION – CONCEPT DEFINITION

Changing the mental image

The most frequent reason for the failure of inventors is an unwillingness or inability to change their preconceived notions about existing ideas. Invention occurs when a concept is altered in some manner. The human mind has the singular ability to grasp broad concepts, but quite frequently it locks inflexibly on a specific concept, unable to conjure new ideas. An inventor must develop an active imagination in order to expand his thinking about a given subject so as to change that subject inventively.

The word image

If ten people were asked to give their first mental image of the word "truck," more than likely each one would have a different picture in mind. For one it might be a tractor-trailer, another a moving van, another a dump truck, and so on. These specific impressions form the basis of the inventive process. They are the foundation stones on which the inventor mentally builds.

A simple aid to inventing is to create new mental images. One way to do this is to take a commonly-known term for a development and consider its synonyms. The mental pictures conveyed by these words will aid in inventing new and related concepts. A thesaurus is a useful tool in this regard. Table 4 below shows a series of synonyms having the broad connotation of eraser.

abrader	eliminator	obliterator	scrubber
blotter	eradicator	polisher	striker
brusher	expunger	purifier	swabber
canceler	hoser	remover	sweeper
cleaner	launderer	scavenger	washer
defacer	mopper	scourer	wiper

TABLE 4. SYNONYMS FOR ERASER

By applying each of these words to the blackboard-eraser concept, you can broaden your inventive perspective and perceive other types of erasers that might perform the same function. These synonyms aid you in devising other approaches to the problem of blackboard cleaning, thus enabling you to invent new devices.

Example 25. The words "eradicator", "launderer", "scrubber", "washer", "scourer", "hoser", "mopper", "swabber", and "cleaner" imply the use of a liquid. Thus you can alter the present eraser by incorporating liquid into it. Figure 26 shows an eraser with a fluid receptacle and means for applying the liquid to the board.

When the fluid release is pushed down, the valves open, allowing the liquid in the reservoir to be discharged, moistening the felt wiper.

Example 26. The word eradicator brings to mind liquid ink eradicator. A liquid could be used to chemically cause the chalk material to disappear. You might also create a new chalk material to work with a special eradicator.

The words "obliterator," "expunger," and "eliminator" have a different implication from those just mentioned. With these words in mind you might consider using a type of gun which can be aimed to clean the board. Figure 27 shows a device that sweeps the surface of the board.

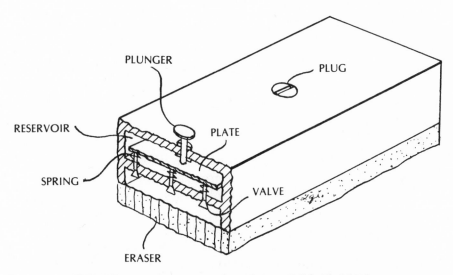

FIGURE 26.   Cut-away of an Eraser With Fluid Dispenser

The chalk is applied and held to the board by an electrostatic charge. The gun removes the chalk by canceling or reversing the charge so as to repel the particles of chalk from the board causing them to fall into the tray.

Other developments may be considered feasible by a study of the words "abrader," "defacer," and "scourer." Thus you might consider a gun that shoots abrasive particles at the board to remove the chalk. The particles can be caught by a vacuum at the head of the gun and the particles recycled to the gun for reuse. A careful study of each new term may develop a new solution to a specific problem.

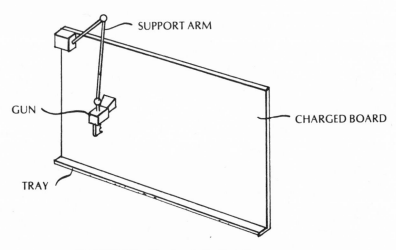

FIGURE 27.   Electrostatic Gun Eraser

The definition image

Another approach to inventing is to write down a definition of the subject you wish to improve. A written definition will permit you to interrelate combinations of words to produce new ideas.

Example 27. A pack of book matches is:

A receptacle for matches including a cover, a striker, a series of connected matches, and means for attaching the matches to the cover.

After defining the invention, take each major term and substitute synonyms in various combinations. Three are used for each major term in Table 5 for purposes of clarity but obviously many more combinations are available.

A	receptacle	for	matches	including	
	container		igniters		
	box		caps		
	holder		wicks		

a	cover,	a	striker,		
	housing		hitter		
	envelope		beater		
	closure		igniter		

a	series	of	connected	detachable	matches,
	plurality		attached	removable	igniters
	group		joined	separable	caps
	strip		united	disconnectable	wicks

and	means for attaching	the matches	to the cover.
	holder	igniters	housing
	staple	caps	envelope
	fastener	wicks	closure

TABLE 5
SENTENCE DEFINITION OF MATCHBOOK WITH VARIABLES

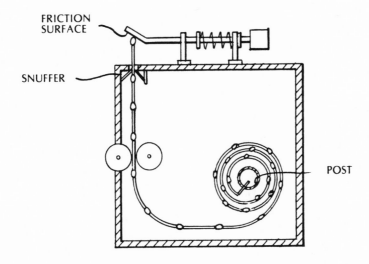

FIGURE 28. Matchbox With Igniter Strip

By using the substitute words in the definition you can create a new variation of an existing idea or a new concept. Thus you can develop from the above:

A **box** of **caps** including a **housing**, an **igniter**, a **strip** of **joined separable caps** and a **holder** for the **caps**.

TABLE 6. NEW SENTENCE DEFINITION OF MATCHBOOK

Figure 28 shows a device fitting the above new description set out in Table 6.

Instead of separate matches we have a strip of paper with a series of spaced igniter areas similar to ordinary caps used in cap pistols. A spring plunger ignites the projecting portion. A new light can be fed by a wheel as needed. The strip is mounted on a post and a snuffer acts as a safety feature to prevent succeeding matches from igniting within the box.

Other word combinations can be devised to aid in creating a new invention. Try each possible combination of words to create new ideas. You will be able to develop a number of different and unique ideas in this manner.

A primary rule to inventing is: Always try the **concept definition** *approach. Use of* **the word** *and* **definition images** *will prove invaluable tools in devising new ideas.*

No. 621,195.

Patented Mar. 14, 1899.

FERDINAND GRAF ZEPPELIN.
NAVIGABLE BALLOON.
(Application filed Dec. 29, 1897.)

(No Model.)

4 Sheets—Sheet 1.

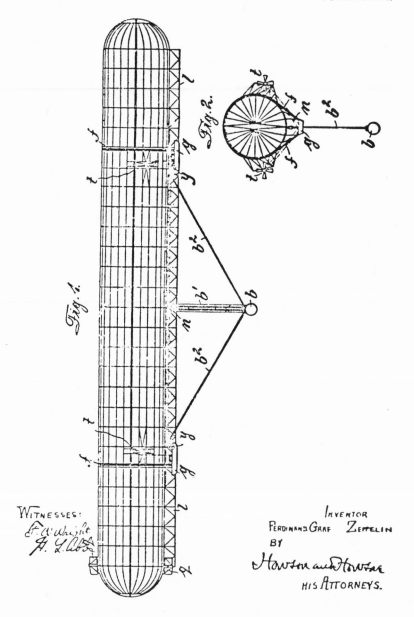

WITNESSES:
F. A. Wright
H. L. Abbott

INVENTOR
FERDINAND GRAF ZEPPELIN
BY
Howson and Howson
HIS ATTORNEYS.

8 / IMAGINATION – COMBINATION AND SUBSTITUTION

Different approaches

No single path leads to a new invention. An inventor must try every avenue possible, much like a chess game, considering all the possibilities, with the hope that eventually one or more will lead to a solution. Although previous chapters illustrated combination and substitution inventions, those inventions were arrived at through other processes. This chapter develops the concepts of combination and substitution.

Combination

Combination takes two or more distinct inventions and puts them together to produce a new invention with each dependent on the other. The following show various approaches to combination:

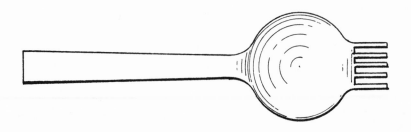

FIGURE 29.   Combination Fork and Spoon

Example 28. A spoon has a function separate from a fork. A spoon is primarily a scoop with a handle on it. A fork is primarily a spearing implement with a handle on it. Figure 29 shows a different combination of these two instruments.

The tines of the fork are short to permit full insertion into the mouth. The bowl portion of the spoon is wide to allow for sipping from the side.

Example 29. The spoon and fork at opposite ends of the handle, Figure 30, provide a dual function but one must frequently reverse the grip on the implement.

Example 30. A popular piece of camping gear has the knife, fork and spoon riveted at the handle ends, pivotable to allow separate use of each implement. This is a typical combination invention. Although the knife and spoon have different functions, the two functions can be combined as illustrated in Figure 31 A combination knife and spoon implement would require care in eating to avoid cutting one's mouth. Thus the knife edge should be on the right side for right-handed people and on the left side for left-handed people.

Example 31. The matchbook can be combined with many other inventions. Figure 32 shows the incorporation of a small mirror on the inside flap.

FIGURE 30.   Combination Fork and Spoon

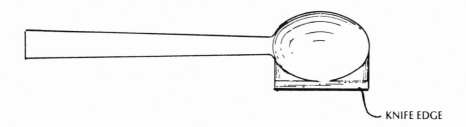

KNIFE EDGE

FIGURE 31.   Combination Knife and Spoon

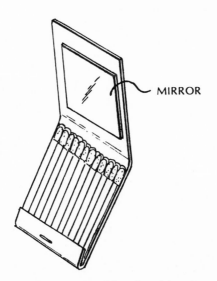

MIRROR

FIGURE 32. Matchbook With Mirror

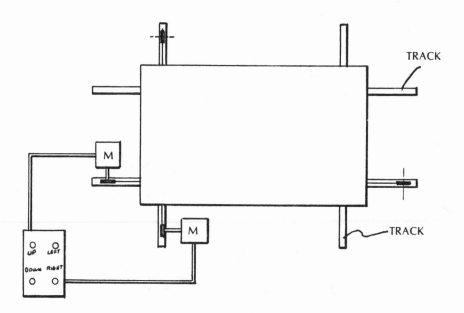

FIGURE 33   Horizontally and Vertically Shifting Blackboard

Example 32. Figure 33 shows a blackboard that shifts right and left or up and down depending on the reach of the teacher using the board. By operating the foot control, the teacher can shift the board like a typewriter so that it is unnecessary to move while writing on the board. In this invention we have adopted the carriage shifting features of typewriters to give versatility to the blackboard. The carriage mechanism may be mounted in the wall.

Example 33. In Figure 34 the simple combination chalk-holder prevents the hands from picking up the chalk dust.

Substitution

Substitution involves replacing one analogous or non-analogous part, unit, system or the like for another to develop a new invention. More than one substitution may be made in a single development so that each original part of a known subject may be substituted for something else.

Example 34. In Figure 35 the lollipop is substituted for the match to produce a lollipop matchbook.

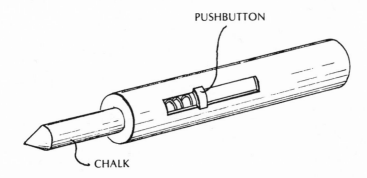

FIGURE 34.   Chalk-holder Implement

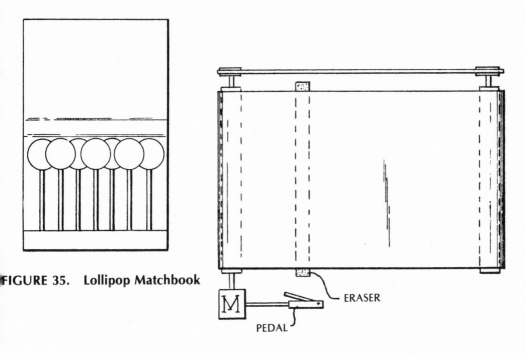

FIGURE 35.   Lollipop Matchbook

FIGURE 36.   Endless Belt Blackboard

Example 35. Other items may be substituted for matches, such as toothpicks, needles, or the like.

In Figure 36, an endless belt provides additional writing surface. An eraser can also be combined in the mechanism to wipe the chalk from the movable belt.

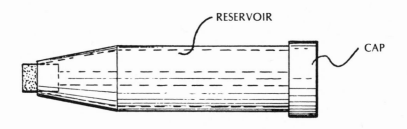

FIGURE 37.   Liquid Chalk Pen

Example 36. Figure 37 shows the replacement of chalk with a felt tip pen having liquid chalk.

The writing pen eliminates the dust problem common with hand-held chalk.

You should study each area of development and try various combinations and substitutions in much the same manner as illustrated. The possibilities for combination and substitution inventions are unlimited.

9 / IMAGINATION – ADDITION, DELETION AND REARRANGEMENT

Addition

Frequently an invention will involve joining similar units. This technique is called addition and is similar to the combination inventions considered in Chapter 8 which join dissimilar units. Examples 38 through 40 are typical.

Example 38. Taking a number of matches and uniting them in strip form is an addition invention. Figure 38 shows an eraser capable of covering a large area. The three ordinary erasers are interconnected by a clamping bracket having a support handle. With this device the board can be erased easily and quickly.

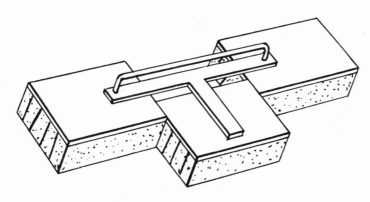

FIGURE 38.   Three Unit Eraser

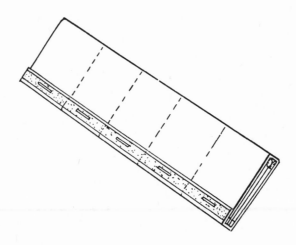

FIGURE 39.   Tear-Off Book Match Strip

Example 39. A scored strip of matchbooks, Figure 39, may be broken away in one or more units as desired.

Example 40. Figures 40, 41 and 42 show the result of adding blackboard surfaces to increase the writing area.

Figure 40 shows a reversible blackboard with front and back writing surfaces, Figure 41 shows a type of blackboard with three revolving, interchangeable surfaces, and Figure 42 shows a pocket blackboard a portion which is extendable when desired. When the board is returned into the recess, a wiper could be provided in the pocket to clean the extendable surface. The wiper could be made inoperative so that writing on the board can be stored for future reference.

Deletion

Occasionally an invention can be created by the elimination of a part or portion of an item. Simplification is an approach an inventor should always consider. Examples 43 and 44 pertain to deletion inventions.

Example 41. Figure 43 shows the center of the eraser removed to lighten the eraser and yet give it the same effective wiping area. This invention provides two cleaning surfaces, one behind the other. The second surface removes dust that the first fails to pick up.

Example 42. Figure 44 shows a variation of this concept.

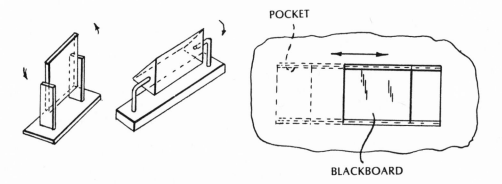

POCKET

BLACKBOARD

FIGURE 40.　　　**FIGURE 41.**　　　　　　　**FIGURE 42.**
Blackboards Having More Than One Face

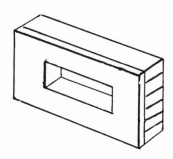

FIGURE 43. Eraser With Center Removed

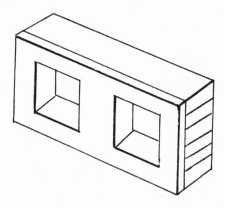

FIGURE 44. Eraser With Two Windows

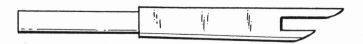

FIGURE 45.   Knife With Fork End

The type of eraser shown in Figures 43 and 44 are much more easily cleaned because the dust travels outward through the openings when the erasers are clapped.

Example 43. A knife having a portion removed to form a fork as shown in Figure 45, makes the knife more useful as a spearing implement. Deletion in this instance created a dual-purpose tool.

Rearrangement – generally

You may also create an invention by shifting, reversing or otherwise rearranging parts as illustrated in Examples 44 and 45.

Example 44. Figure 46 shows a knife with a removable handle that can be secured to the blade in two different positions; position **A** produces a standard knife; position **B** produces a chopper.

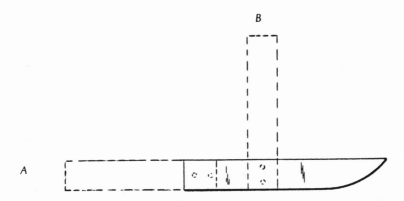

FIGURE 46. Knife With Two-Position Handle

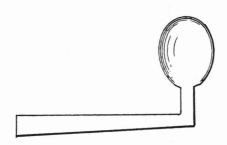

FIGURE 47. Right-Angled Spoon

Example 45. The bowl and handle of a spoon in Figure 47, are at a right angle for convenient scooping with a wrist action.

Rearrangement – adjustability

The concept of rearrangement includes the idea of adjustment. Adjustment includes fixing, setting or holding a selected position. Thus, an adjustable device normally has some mechanism to hold the selected adjustment by friction, set screws, pins, clamps, and the like.

Examples 46 to 49 show the use of various adjustment techniques.

Example 46. An adjustable eraser is shown in Figure 48 for wiping small or large areas. The adjustment length is held by the friction fit between adjacent felt strips.

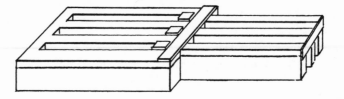

FIGURE 48. Adjustable-Length Eraser

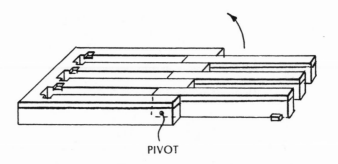

PIVOT

FIGURE 49. Foldable Eraser

Example 47. It is to be noted that the construction of the eraser in Figure 48 allows for an infinite number of wiping adjustments between the closed and the extended positions. The hinged eraser in Figure 49 allows only two effective wiping positions: folded and unfolded.

Example 48. The erasers shown in Figures 48 and 49 may not include the best adjustable arrangement. Figure 50 shows another possibility that affords longitudinal as well as lateral adjustment.

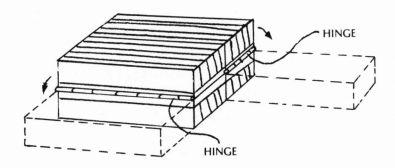

FIGURE 50. **Pivotally-Adjustable Eraser**

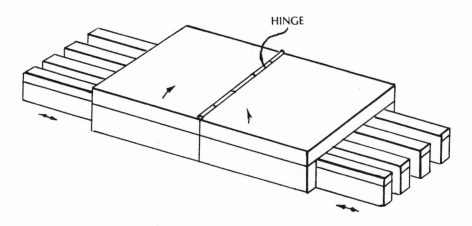

FIGURE 51. **Dual-Adjustment Eraser**

Example 49. Although the eraser in Figure 50 has a greater degree of adjustability than the two previous Figures, Figure 51 shows an even greater adjustability in that it has pivotal as well as sliding adjustability.

Adjustment may be in increments as illustrated in the previous Figures. These important aspects are sometimes determined by such factors as assembly, economics, need and the like.

Study adjustability carefully to assure maximum advantages of this inventive tool.

Addition, deletion and rearrangement are important considerations and you must keep these in your check list of invention steps.

G. WESTINGHOUSE, Jr.
STEAM POWER BRAKE.

No. 88,929.

Patented Apr. 13, 1869.

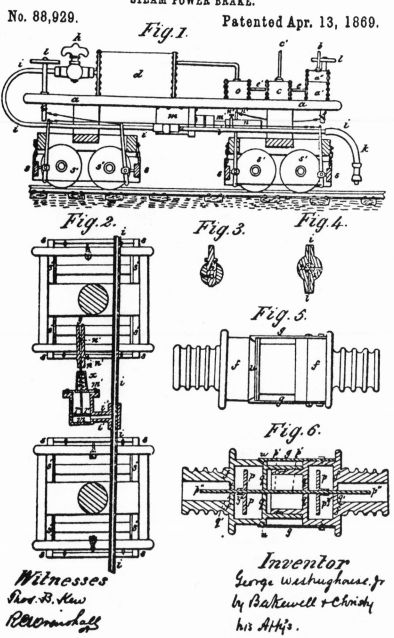

Fig. 1.

Fig. 2.

Fig. 3.

Fig. 4.

Fig. 5.

Fig. 6.

Witnesses
Thos. B. Kerr
R. Crandall

Inventor
George Westinghouse, Jr.
by Bakewell & Christy
his Atty's.

10 / IMAGINATION – PHYSICAL FORCE AND EFFECT

Change

The application of a physical force to a subject can effect inventive change in the subject. The change may be the result of a critical force. The determination of this critical force is frequently the result of extensive experimentation. This is particularly true in the fields of chemistry and electricity. This chapter deals generally with such forces and effects.

Pressure

There are many ways in which the application of pressure produces invention. Table 7 shows typical examples.

abrading	drawing	milling	springing
absorbing	drilling	peening	squirting
beating	exhausting	polishing	spinning
bending	expanding	pounding	stamping
blowing	exploding	pressing	stirring
bonding	extruding	pulling	stretching
cauterizing	flexing	ripping	stripping
centrifuging	fluidizing	rubbing	sweating
chewing	forcing	rotating	tearing
clipping	forming	routing	tensioning
clinching	gasing	standing	turning
comminuting	grasping	shearing	twisting
compressing	gripping	shooting	vacuuming
crushing	hydraulicking	shredding	whipping
decordicating	inflating	squeezing	wrenching
deflating	kneading	splitting	vulcanizing

TABLE 7. LIST OF PRESSURE FORCES

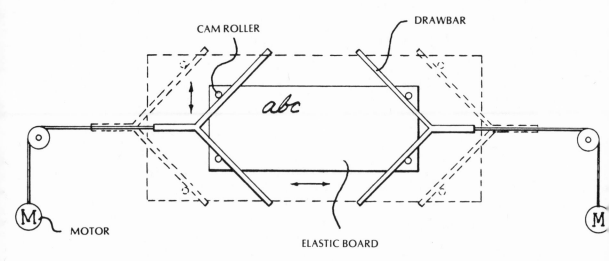

CAM ROLLER

DRAWBAR

abc

MOTOR

ELASTIC BOARD

FIGURE 52.   Stretchable Blackboard

Since this is not a complete list of all pressure forces, you should add to the list when a new force becomes known. The list should be reviewed whenever you are working in a new area of development on the chance that one of the pressure forces might suggest a new approach to an invention.

Example 50. A typical application of a pressure force selected from the list as applied to a blackboard is shown in Figure 52.

Small letters are written on the blackboard while it is in the relaxed or non-stretched position. When the blackboard is stretched, the letters enlarge enabling one to read from a greater distance away. The cam rollers move laterally away from each other on the "V" drawbar stretching the elastic board vertically and horizontally.

Example 51. Figure 53 shows another application of a pressure force to an eraser. The sponge portion of the eraser can be squeezed to expel absorbed liquids.

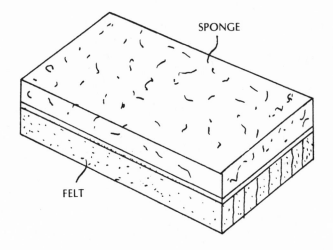

FIGURE 53. **Eraser With Sponge Backing**

Temperature

The application of heat or cold can produce new and inventive devices and processes. In the area of electricity and chemistry particularly, changes in temperature produce new developments. A heated blackboard could change the color of a special temperature-sensitive chalk. The color change would occur only on the application of heat. Thus chalk applied to a surface would have any of a range of colors depending on the temperature of the board. It would also be possible to write on a cold surface, heat the surface to change the color, let it cool, and write again on the cold surface to have a dual-color system. Several different color changes could occur at different temperatures to permit other colors to be included in the system.

Radiation

The application of radiation produces many invention possibilities as we have already noted in earlier chapters. In considering radiation principles, you should have a checklist such as in Table 8:

alpha rays	laser light
beta rays	phosphorescent
cosmic rays	radio waves
fluorescent	ultra violet light
gamma rays	white light
infra red light	x-rays

TABLE 8. TYPES OF RADIATION

You should add to the above list as new radiation effects are developed.

Gravity and such forces

Table 9 should also be considered as possibly leading to new inventions:

acceleration	gravity (weight)
centripetal	oscillation
centrifugal	pendulum
deceleration	reciprocation

TABLE 9. LIST OF FORCES

You might note some overlap between lists. If a question arises with a listing, then add to each of the lists to keep current with new developments.

Conductivity and magnetism

Many inventions involve the use of conductive or magnetic principles. Conductivity includes heat, light, electricity and sound.

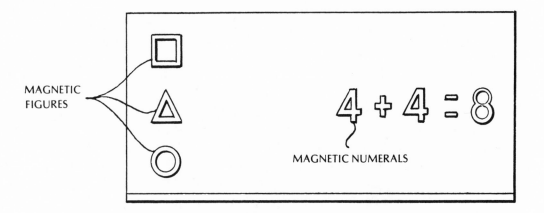

FIGURE 54. Steel Blackboard

Example 52. Figure 54 shows a coated steel blackboard on which magnetic letters, numbers and figures are held in place by magnets. Thus a teacher could use pre-designed figures to eliminate much actual sketching.

Miscellaneous characteristics

In addition to the forces and effects previously mentioned, consider the following list of other various forces and effects as possibly causing invention.

buoyancy	flexibility
brittleness	frangibility
crystallinity	friability
density	hardness
ductility	maleability
	permeability

TABLE 10. MISCELLANEOUS CHARACTERISTICS

Consider each of these tables as a possible new cause or effect in a given development area. Many developments are improved by a change or application of a particular characteristic of the material, article, or machine to include or exclude one or more other physical forces or effects. You should be alert for new forces and effects to add to the various tables set out in order to expand your inventive potential.

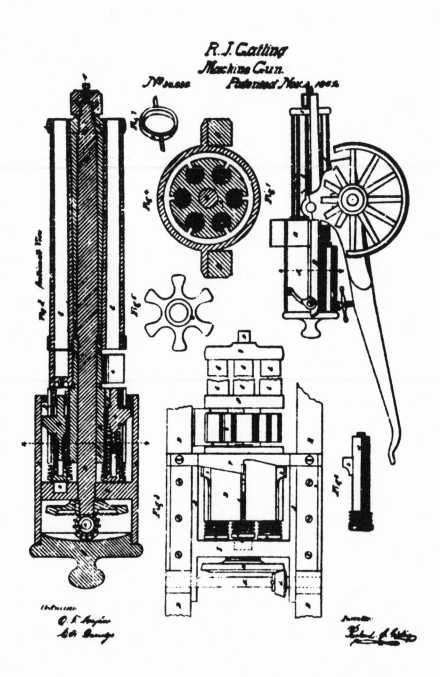

11 / LIMITATIONS— THE CONDITIONS IMPOSED ON DEVELOPMENT

Limiting factors

A thorough knowledge of the laws, controls, restrictions, regulations, requirements and restraints existing with regard to a specific development area can be an important aid to an inventor. If you are aware of the parameters (limitations) of your field of endeavor, you will be less likely to create something having little utility. A builder would not construct a home without knowing the building code regulations which tell him what is and what is not permitted. These regulations force him to build to accepted standards.

Regulations are constantly changing. For the inventor, these changes create new problems for which solutions must be found. An inventor must therefore keep abreast of new developments, new materials, and new methods in his field to assist in solving these newly created problems.

The following considerations should be carefully reviewed on a periodic basis to determine whether or not effort should be expended to devise new machines, articles and processes to meet the changes in conditions:

1. Materials
2. Equipment and Labor
3. Time
4. Space
5. Storage
6. Assembly and Disassembly
7. Use
8. Transportation
9. Safety
10. Precision

Materials

You should always keep in mind regulations concerning the materials ordinarily used. Many inventions fail or have limited success because the inventor disregarded material characteristics. Automobile recalls are frequently the result of poor selection of materials. A more careful choice would prevent expensive recalls.

You should also check the availability of the materials to be used. In many instances, various materials will work in a particular application. Selection should be based on such factors as durability, suitability, availability and cost. Since new materials are constantly being developed, it is important that you study developments in new materials to determine if any can be substituted for present materials.

Nylon ® and Teflon ® are typical examples of materials that were found to have many unexpected uses resulting in thousands of new inventions ranging from carpets to nonsticking frying pans.

Blackboards were first made from slate. Today, man-made compositions serve the same purpose less expensively and more durably.

Wood veneer is a development which replaced solid wood because of the scarcity of fine woods. New developments in veneering have created more durable and less expensive pieces of furniture and are as beautiful as the solid woods.

Although paper is the accepted material for the bookmatch, other materials may reduce cost and produce a better, longer-lasting light You might consider this problem in detail in view of the many materials available.

Equipment and labor

The availability and cost of machinery and labor are among the chief factors in the development of new machines and techniques. The farmer is hard pressed to find labor today. One hundred years ago, a farmer could barely handle more than 40 acres by himself. But today, advanced agricultural machines do the work of many field hands, enabling a single farmer to work hundreds of acres.

Laws governing labor and equipment cause new inventions to occur because of the need for better and safer machinery. Laws also outdate machinery. If the law requires special equipment, keep in mind that modification of outdated equipment may make it perform better or do a job that it was not initially designed to do. Dual-function machines frequently result from new combinations of old equipment designs.

Time

Time regulation and restrictions are frequently imposed on invention. Time is a commodity; it commands a price. To save time, you may modify and improve existing developments. In general, time is an aspect of all invention. Study how long an item will operate before it breaks down or ceases to function. You must consider time and expenditures. To save time, you should carefully study continuity in flow of operation and time intervals between successive operations.

Careful analysis of these studies may suggest new inventions. Heavy traffic or obstacles in a plant or shop can be disruptive, causing time delays. Time study of production limitations may show how procedures can be performed in a shorter period by changing a work system, removing obstacles, or devising a machine having a higher rate of speed.

In devising continuous-flow-systems, and machines or devices, you should ask yourself, "How long does this now take?" and "Can it be done faster?".

If production lines can be shortened, time and space can be saved. Regulations on time should be considered well in advance of delivery date. This enables one to install special techniques and equipment in order to meet requirements.

FIGURE 55. 4-Step Process

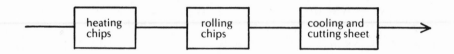

FIGURE 56. 3-Step Process

Example 53. The production line operation of Figure 55 shows a method of making plastic sheet material. The plastic chips are heated and rolled to produce sheet. The sheet is then cooled and subsequently cut. Figure 56 shows a process in which the cutting and cooling stages are combined. This can be done by lubricating the knife to avoid sticking. Combining the two steps reduces the length of the production line.

Another area of time involves the extension of the life of an operation, function, or item. Extending the time period results in many new inventions.

Example 54. Thus you might find a need for a long burning match which will give progressively greater heat and light.

Figure 57 shows a match which will produce a larger flame as the match burns. The conical shape of the base causes a gradual slowing down of the burn.

Example 55. To produce a fast-burning match, we might make the match hollow to produce a chimney effect as noted in Figure 58.

Lengthening the duration of operation should be considered in any development study.

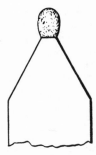 

FIGURE 57. Slow-Burning Match   **FIGURE 58. Fast-Burning Match**

Space

You should give thought to the type of space requirements a device must have. These requirements may apply to the table below:

use	assembling
storage	disassembling
packaging	safety
shipping	

TABLE 11.   SPACE REQUIREMENT CONSIDERATIONS

The design of a piece of equipment should include considerations as to all of the above. If an item has infrequent use, it may be stored in a more remote place than one having frequent use. Location of bolts, screws, nuts, holddowns, other fasteners, and the like are important particularly if space is at a premium and assembly and disassembly are considerations. Furthermore, space requirements may necessitate the design of a special tool to work on a new development. Thus, the automobile engine presents many problems to the mechanic and the designer should consider location of equipment beneath the hood not only as to ease in removability of certain parts but also use of tools.

Space requirements must be considered with regard to packaging and shipping. Some items require special containers for careful handling. You must allow cushioning space when designing a package for shipping. Remember also that inadequate space may create a hazardous condition. You must anticipate and design so that hazards will be minimized. You should also consider storage problems before finalizing an invention. If an invention becomes too difficult to store because of size or some other reason, advanced design may overcome these particular problems.

Stacking features are important; consider devices and techniques for stacking such as rims, bumps, lips and the like which permit easy shipping and storage with minimum damage though the articles are stacked one on the other. Stacking features also help to reduce space requirements.

Storage

Shelf life relates to the length of time an item can be stored before it deteriorates. You should always determine the shelf life of any new item since its value and utility may be enhanced. Such an improvement may in itself be inventive and it may occur because of a change of materials, packaging, and the like.

Many other conditions are imposed on an item in storage. Table 12 lists some of the more important conditions:

time	explosive conditions
temperature	humidity
light exposure	composition
effective electrical fields or magnets	pressure changes
size of the storage facility	

TABLE 12. CONDITIONS AFFECTING STORAGE

Assembly and disassembly

Study whether a machine, device or system must frequently be taken apart and reassembled. Your findings may become the basis for an important machine, system, part or the like. Design new equipment with these thoughts in mind.

FIGURE 59.   Ring Post Connector   **FIGURE 60.   Spade Post Connector**

FIGURE 61.   Locking Spade Post Connector

Example 56 and 57. Two different type of connectors for electrical posts are shown below in Figures 59 and 60.

In Figure 59 the connector is not easily applied or removed from the posts since the nut must be fully taken off before it can be removed. The advantage is that it is built to stay on even if the nut becomes loose. The connector of Figure 60 is easily applied and removed from the post by merely loosening the nut but the connector will not stay on if the nut is loosened.

Example 58. By combining the features of the two devices shown in Figures 59 and 60, we can produce a connector such as shown in Figure 61.

Note that the post must be somewhat larger than the opening between the arms of the connector making it difficult to remove even if the nut is loosened. Thus it will stay on the post in the same manner as the ring connector if the nut becomes loosened, but it can be removed by applying enough pressure to snap the end of the arms around the post once the nut has been loosened.

Use and non-use

Inventions occur involving expendability. A throw-away device may be acceptable in many instances where labor is a problem. You must consider that many developments are subject to frequent or infrequent operation. Those involving infrequent operation are generally less expensive and require less maintenance than those involving frequent operation. Remember you can overbuild without any substantial gain.

In general the following Table 13 lists some of the many things you should give thought to when studying particular problems of use and non-use.

permanency	radiation
dust and dirt	magnetic forces
weather	light
temperature	wind
pressure	precision
gravity	

TABLE 13. ITEMS AFFECTING USE AND NON-USE

Other aspects of use and non-use may occur to you when you are developing an idea. Add them to the above list for a ready reference.

Transportation

Regulations or requirements involving the transportation of a development should be checked in advance with the idea of possibly improving the item. The following Table 14 may give you an indication of the areas which could be improved:
vibration
temperature
humidity
pressure
electrical and magnetic problems
light

TABLE 14.   CONDITIONS AFFECTING TRANSPORTATION

As in the previous tables, it is well to add to these as new regulations are passed.

Safety

Many laws are designed to protect the worker. Thus, after you have developed an invention, study it for safety aspects and make whatever changes in design may be necessary in order to meet any regulations or standards which might exist or might be predicted.

Precision

Precision is frequently regulated by military government or health requirements and the like. You may fail to make a noteworthy invention if you forget to take into consideration the precision required by the purchaser.

In general, therefore, you should run a careful check of all of the above aspects when inventing and keep tables current for ready reference and adding to them as you discover a new condition.

E. Howe, Jr.

Sewing Machine.

Nº 4750   Patented Sep. 10, 1846.

Fig. 1

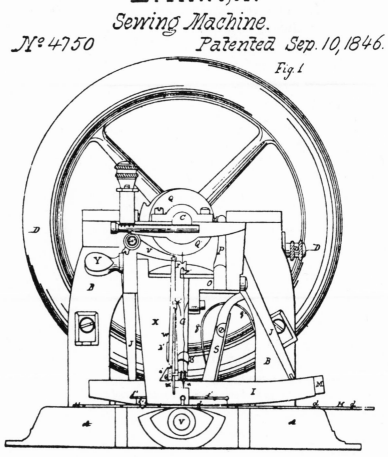

Fig. 4.   *Fig 7*

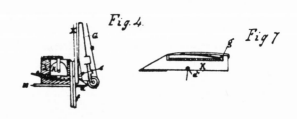

12 / MISCELLANEOUS AIDS IN INVENTING

Brainstorming

Brainstorming is a well known system in which a group of individuals get together for the purpose of combining their talents to work out a problem and arrive at a new invention. Many books have been written on this subject.

Brainstorming is an excellent approach to problem solving encompassing more than one technical area. Two or more technicians each having distinct special skills may contribute their individual specialties to develop a solution. In aerospace technology, for example, medical doctors, nutritionists and chemists would contribute their knowledge to develop special foods or packages for interplanetary flight. The interchange of ideas between different specialities is of the utmost importance in developing an invention which involves more than one technical area.

Some problems are created where the brainstorming technique is used in an area in which the technology is limited to one field only. In these instances, each individual has substantially the same general background and frequently each attempts to outdo the other. This is particularly true when the reward for discovery is great. A more analytical and cautious person is often frustrated by impulsive thinkers. This does not mean that the more analytical man may not develop a better solution to the problem. A cautious man studies all of the possibilities carefully and does not jump to an immediate conclusion. Impulsive men frequently reach conclusions faster but these conclusions are not always well developed. Furthermore, brainstorming can result in jealousies in a research group, creating many problems. It should be used with care and discretion.

The last questions

After thinking of an invention, the inventor should carefully check to determine whether it is likely to succeed. A great deal of money and time are expended on items which eventually turn out to be impractical. Many people rely on the inventor to foresee the practicality of his invention but he is frequently unable to discern the problems of the development either in production, marketing, handling or the like. The inventor should discuss his proposals with a number of specialists and carefully weigh their opinions as to the practicality and likelihood of success, particularly before any large sums of money are expended on items such as prototypes, production charges, advertising and marketing. Cost is not always a factor and an item can be successful even though it is more expensive than another already on the market. Sometimes advertising can successfully sell a more complex and expensive item, particularly because tastes vary. Thus a market exists for many different automobiles though they all basically serve the same function. Many intangibles create a market for a particular item other than pure practicality. However, in general, the inventor should work towards a practical solution to a problem.

Acquisition of knowledge

There are many ways of acquiring knowledge and information. Several have been discussed in earlier chapters; however; whenever possible, you should also consult experts in the field. Gaining information from literature may not always be as valuable as direct contact with a specialist. A conference relative to a specific problem area with an expert might re-solve certain issues faster than reading a half dozen books on the broad subject. In this regard, it is worthwhile for a person working for a corporation to check with other work groups in the corporation to avoid the possibility of duplication.

Subscribe to publications in your specific field in order to keep abreast of new developments. These publications will more than pay for themselves in time.

Since the United States Patent Office is a main source of technical information of a classified nature, you should use this facility and even discuss problems with the various exam-iners who, as experts in specific fields, can be very helpful. The Patent Office facilities are unequaled in volume and scope of technical information. Certain corporations and technical as-sociations have highly-specialized libraries dealing with a spe-cific subject and these corporations and associations may also prove to be invaluable for gaining additional background in-formation. For the most part, unless the item is of a secret na-ture, most corporations and associations are quite helpful in orienting an inventor in a certain area and releasing certain technical data and information requested. In addition to the Manual of Classification available and referred to in Chapter 3, many countries in the world have a weekly publication which carries an abstract of each patent issued that week. A subscrip-tion to this publication will keep one abreast of the newer de-velopments as they are patented. More detailed information about the invention can be obtained from a copy of the patent since the abstract in the weekly gazettes does not contain all the detailed information of a patent. Patents can be obtained from the patent offices of various countries in the world for a relatively reasonable sum.

An inventor should not be easily discouraged. He should have tenacity and stick with the problem even though it might appear insurmountable. The more information he acquires in a specific area, the closer he gets to the solution. An inventor should develop a certain amount of self assurance. An in-ventor should not be in a hurry to arrive at a conclusion. He should be objective, but ready to change his approach and make broad evaluations without being unnecessarily influ-enced by others.

An inventor should set aside a specific time for inventing and do it on a routine basis so as to develop a pattern of think-ing. Some inventors may prefer to work in the morning, while others prefer to work at night. A pattern enables an inventor to review old ideas and to look at them in a new light. Be persistent but cautious in research, realizing that false information can create confusion and delay the discovery of an invention. If doubts arise as to information given by experts, compare the information given with other sources to confirm its reliability.

As a last question, an inventor who has developed a new item should ask, "Can something else take its place?" He may, by asking the question, come up with a new item. Inventions in general are made by someone who has determined that something can replace an earlier invention.

13 / SUMMARY OF INVENTION STEPS

Review

Having now considered in detail the various approaches which may lead to invention, this chapter systematically summarizes the step-by-step processes with the following providing a ready reference and reminder.

Inventive steps

1. Look for the complaints which develop in the problem area and seek to devise methods for overcoming the complaints as explained in Chapter 2, **Problems – complaints.**

2. Look for abnormal situations, recurring breakdowns and injuries and seek to devise methods of overcoming these problems as explained in Chapter 2, **Problem – difficult inconvenient, and abnormal situations, recurring breakdowns – injuries.**

3. Construct tables showing the historical development of the problem area similar to Tables 2 and 3 in Chapter 3, **Inventing through history.** The tables will help evaluate the practicality of the invention and the possible future economic growth and need; they will also aid in projecting trends and assist in planning future research and development projects.

4. Classify the problem area in accordance with a known classification system such as shown in Plates 1 through 6 and in Chapter 3, **Inventing through classification.** The classification tables should give you many ideas in your problem area and rapidly bring you up-to-date on similar areas of development.

5. Make a detailed qualitative and quantitative analysis of the problem area to answer the question **DATA — WHAT IS IT?** as set out in Chapter 4, and seek to devise ways to change the subject area by using different compositions changing sizes, shapes, colors, weights and other characteristics.

6. Use **The Tree System** approach on the problem to answer the question **DATA — WHY DOES IT EXIST?** as described in Chapter 5 and consider the primary objects of the subject area and its advantages and disadvantages, including the effect on the five senses and seek to devise ways to improve the advantages and eliminate the disadvantages.

7. Determine **How and why does it work?** as set out in Chapter 6, and endeavor to improve the functional aspects of the problem area.

8. Find out the exact time of use of a subject and inquire: "When is it used? and When can't it be used" as set out in Chapter 6, **When is it used?** and devise means for making it more useful.

9. Determine where the subject is used and where it cannot be used as described in Chapter 6, **Where is it used?** and devise means for giving the development greater utility such as making it operable in areas where previously inoperable.

10. Determine the system and manner of use of a subject as described in Chapter 6, **How is it used?** and endeavor to improve the techniques or methods of using the system by adapting it to other systems. Consider **Who will be the user?**

11. Define the subject by creating a synonym table similar to Table 4, Chapter 7, **The word image** and apply synonyms to the subject concept to create new approaches and new devices.

12. Define the subject by creating a sentence definition as shown in Example 27, Chapter 7, **The definition image** and substitute words or phrases for the primary definition words or phrases as shown in Table 5, Chapter 7, and endeavor to devise new variations of the existing concepts by interchanging words or phrases in the sentence definition.

13. Combine two or more different subjects one or more of which is in the subject area as set out in Chapter 8, **Combination,** to create new combination inventions.

14. Substitute analagous or non-analagous parts, units or systems; for existing parts, units or systems as set out in Chapter 8, **Substitution** to create new substitution inventions.

15. Add similar units to an existing subject in the problem

area as set out in Chapter 9, **Addition,** to create multi-effect devices.

16. Delete parts or portions of the subject as set out in Chapter 9, **Deletion,** to create new developments.

17. Rearrange parts or systems in a problem area as set out in Chapter 9, **Rearrangement** and consider adjustability as it applies to the problem area to arrive at new inventions.

18. Consider the effect of physical forces on the basic problems, such as pressure, temperature, radiation, gravity, conductivity, magnetism and the like as set out in Chapter 10, **IMAGINATION – PHYSICAL FORCE AND EFFECT** and use tables where possible. Review the tables applying each of the listings to the subject area to arrive at other operable methods and devices capable of performing the same or similar functions or results.

19. Study all of the limiting factors restricting the area of invention including the regulations, laws, controls, requirements, and restraints as discussed in Chapter 11, **LIMITATIONS – THE CONDITIONS IMPOSED ON DEVELOPMENT.** Consider these factors carefully and compile tables for quick review whenever possible. This may lead to noteworthy improvements in the subject under research well in advance of the imposition of a new regulation.

20. Consider brainstorming as discussed in Chapter 12, **Brainstorming,** as a possible means of exploring invention, particularly where different technical areas are interrelated.

21. After you have taken all of the above steps, review the chapters in detail and recheck each of the above steps. Sometimes a review after letting the matter rest for a while will be helpful in producing ideas, since it gives the mind time to mull over the finer points of the problem.

Conclusion

Each of the above steps is a key. It could open the door to a new invention. Invention is a matter of trying each key on your particular problem while systematically assembling all the knowledge you have in an endeavor to arrive at a solution that is novel and useful. By using this step-by-step approach you will find that you can invent. Practice on simple items at first and you will be able to work into more complex problems. Remember the world advances with new ideas. Why, therefore, shouldn't yours be among them to aid its advancement?

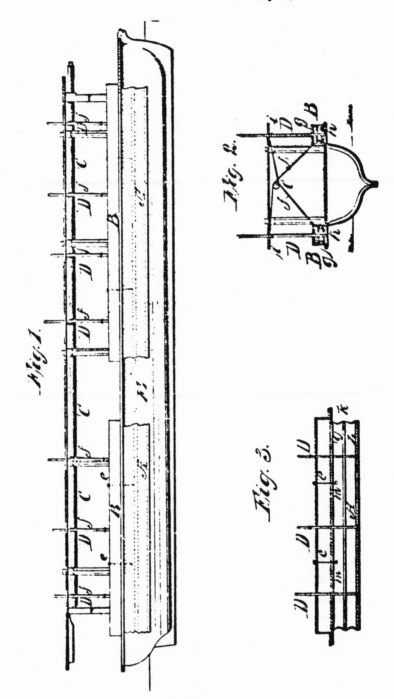

Fig. 1.

Fig. 2.

Fig. 3.

14 / THE HISTORY OF INVENTION IN THE UNITED STATES

Non-patented inventions

The history of invention in the United States is a long one. For the most part, inventions have been associated with patents, but it is to be noted that many advancements in the United States have been made by inventions which were not patented. The formula for "COCA COLA®" has never been patented. It is held as a trade secret. Only a small percentage of the thousands of ideas that are submitted to industry each year by engineers, technicians and research people are patented. The ideas submitted are generally reviewed by a corporation and weighed as to the need for patent protection. For the most part, corporations decide to apply for patents when the invention is such that if not patented, it would be used by competitive companies. If the invention is practical only in the specific corporation environment, there may be no actual need for obtaining a patent. Economics, advertising and goodwill are all factors which corporations consider when selecting inventions for filing before the Patent Office. These factors also govern the independent inventor's decision to file for a patent. Though there are many non-patented inventions, it is realistic to state that the industrial might of the United States is based primarily on patented inventions.

Early patents

The first patent on the North American Continent was granted to Samuel Winslow by the Massachusetts General Court in 1641 for a novel method of making salt. In 1646, Joseph Jenkes received the first patent on machinery from the same court for a mill for manufacturing scythes.

Many of the colonies granted patents but there were no general laws providing for the granting of patents and in each instance, it was necessary for the inventor to make a special appeal to the governing body of his colony.

The problem of getting protection to inventors and authors was considered in framing the Constitution. And thus, article 1, section 8 of the Constitution provided:

"Congress shall have power * * * to promote the progress of science and useful arts by securing for limited times to authors and inventors the exclusive right to their respective writings and discoveries."

On April 10, 1790, George Washington signed the Act which laid the foundation of the modern U.S. patent system. Of the patent system, Thomas Jefferson stated:

"The issue of patents for new discoveries has given a spring to invention beyond my conception."

Abraham Lincoln said:

"The patent system added the fuel of interest to the fire of genius."

Under the patent system American industry has flourished and new products invented which have given employment to millions. The U.S. patent system enabled a small struggling nation to grow into the greatest industrial power on earth.

When George Washington signed the Patent Act of April 10, 1790, for the first time in history, the intrinsic right of an inventor to profit from his invention was recognized by law. Prior to this time, privileges granted to an inventor were dependent upon the prerogative of a Monarch or upon a special act of a legislature. Thomas Jefferson, then Secretary of State, was in effect the first administrator of the American patent system since his department was assigned the task of administering the laws. Jefferson became the first patent "examiner". It was he who made the initial decision to grant or not to grant the first patents. From personal records, it appears that he made an examination of all applications for patents that came before the board set up for granting patents. Jefferson was a mathematician, astronomer, architect, student of languages and a most accomplished and versatile man. He appreciated the value of patents from an intellectual viewpoint, but more because he himself was an inventor, although he never took out a patent on any of his inventions. One of his inventions dealt with the improvement in the moldboard of the plough. This invention had a significant effect on the agricultural development of this country and earned him an award from the French Institute. He also invented a revolving chair which his enemies accused him of designing "so as to look all ways at once", a folding chair or stool which could be used as a walking stick, a machine for treating hemp, a pedometer and a clock system. Because of his appreciation of invention, when he was minister to France, he continually sent back news of the latest European scientific developments and he was the first to notify this country of James Watt's development of the steam engine. It was therefore fitting that he should have been selected to administer the first American patent laws.

On July 31, 1790, the first U.S. patent was issued to Samuel Hopkins of Pittsford, Vermont which involved "the making of Potash and Pearl ash". From that time on, patents issued with increased frequency. On March 14, 1794, Eli Whitney, only two years out of Yale, received his patent on the cotton gin. This invention had a vital bearing on the rapid development of the American textile industry.

On May 5, 1809, Mary Kies of Killingly, Connecticut, became the first woman to receive a United States patent. Her invention related to "the Weaving of Straw with Silk or Thread".

On June 21, 1834, Cyrus McCormick of Virginia received his patent on the reaper. This contribution made the vast fields of the West available for the production of grains so necessary for the world's needs. The West was also influenced by the invention of Samuel Colt for which on February 6, 1836, he received a patent on his famous "Six-shooter".

Patents in the mid-1800's

In 1842, a law was enacted making designs patentable, and by 1965 over 200,000 design patents had been issued. The Act of 1842 also required the owner of a patent to mark his article as patented and include the date on which the patent was issued. The Act also provided a penalty for falsely marking articles patented.

On June 15, 1844, the United States Patent Office granted to Charles Goodyear, Patent No. 3,633 for an "Improvement in the Manner of Preparing Fabrics of Caoutchouc or India-Rubber". The Goodyear process of vulcanization gave rise to numerous great industries as also did Elias Howe's Patent No. 4,750 for an "Improvement in Sewing Machines" granted September 10, 1846.

Few people know that our great President Abraham Lincoln was himself an inventor and received United States Patent No. 6,469 for "A Device for Buoying Vessels over Shoals". The invention consisted of a set of bellows attached to the hull of a ship below the water line which could be filled with air to aid in floating the vessel over a sand bar. The model which Lincoln made can be seen at the Smithsonian Institution in Washington. Because Lincoln was himself an inventor and appreciated invention, he was primarily responsible for the development of John Erickson's "Monitor", the ironclad which fought the "Merrimac". Further without Lincoln's insistence, the Spencer Repeating Rifle would never have been adopted by the Army.

It is interesting to note that in 1857 the United States issued 2,910 patents—about 35% more than Great Britain, which had a far larger population. Prussia that year granted only 48 patents and Russia 24. This noteworthy interest in invention by the people of the United States at this early date was a harbinger of America's industrial success.

On March 1861, a law was enacted which provided a term of patent grant of 17 years for utility patents and a maximum of 14 years for design patents.

It is noteworthy that the Confederacy issued 266 patents mostly concerning implements of war, and that the first Confederate patent, issued to Van Houten of Savannah, Georgia, was for a "Breech Loading Gun".

In 1862, Richard J. Gatling received Patent No. 36,836 on one of the first successful machine guns.

With the end of the Civil War, there was a noticeable increase in invention with 10,000 applications for patents being filed in 1865, 15,000 applications in 1866, and 20,000 in 1867.

In 1868, Christopher L. Sholes was granted Patent No. 79,265 for a typewriter which invention was the precursor of many industrial giants.

In 1869, George Westinghouse, Jr. received Patent No. 88,929 for his famous "Westinghouse Air Brake" which made it possible to make up extremely long trains enabling railroads to handle vast traffic with high speed and safety. This single invention aided in revolutionizing rail transportation in the United States and further aided in the development of our industrial might.

Patents in the late 1800's

In 1870, John W. Hyatt, Jr. and Isaiah S. Hyatt, of Albany, New York received Patent No. 105,338 for "Improvements in Treating and Molding Pyroxyline". From this invention, the great celluloid industry and its many thousands of articles developed.

Mark Twain (Samuel L. Clemens), on December 19, 1871, received Patent No. 121,992 for "Improvement in Adjustable and Detachable Straps for Garments". Mark Twain received three patents including his famous "Mark Twain's Self-Pasting Scrapbook", and his game to help players remember important historical dates. The scrapbook made Mark Twain a sizable profit although it was simply a series of blank pages coated with gum or veneer. The sale of 25,000 during the royalty period lead one of Twain's biographers to comment that this "was enough for a book that did not contain a single word that critics could praise or condemn." Twain's character "Sir Boss" in his "Connecticut Yankee at King Arthur's Court" remarked that "a country without a patent office and good patent laws is just a crab and can't travel anyway but sideways and backways."

In 1873, Eli H. Hanney, of Alexandria, Virginia received Patent No. 38,405 for "Car Couplings". The automatic car coupler, together with Westinghouse's air brake, made possible the gigantic railroad industry of the twentieth century, for without the car coupler, the toll of railroad accidents would be appalling.

In 1874, Joseph F. Glidden, of De Kalb, Illinois received Patent No. 57,124, for "Improvement in Wire Fences". This invention known as barbed wire made possible the fencing of Western farmlands and was the death knell of "the open range".

Between 1876 and 1880, the following monumental patents were granted including Patent No. 174,465 on "Telegraphy" to Alexander Graham Bell, Patent No. 200,521 for a "Phonograph or Speaking Machine", and Patent No. 223,898 for "An Electric Lamp for Giving Light by Incandescence" both to Thomas A. Edison, of Menlo Park, New Jersey. These inventions alone changed the entire course of World history.

Between 1880 and 1900, Elihu Thomson received Patent No. 347,140 for "Apparatus for Electrical Welding"; Nikola Tesla received Patent No. 382,230 for "Electrical Transmission of Power" which patent was the genesis of the induction type of electric motor, so widely used in modern industry; Charles M. Hall received Patent No. 400,665 for "Manufacture of Aluminum" which has become indispensable in numerous industries today; Ottmar Mergenthaler received Patent No. 436,532 for "Machine for Producing Linotypes, Type Matrices, etc."; Frederic E. Ives received Patent No. 495,341 for "Photogravure Printing Plate"; Whitcomb L. Judson received Patent No. 504,038 for the Slide Fastener now commonly known as the Zipper; Edward G. Acheson received Patent No. 560,291 for the "Electrical Furnace" making possible the production of carborundum, which is one of the hardest substances known and widely employed in industry for cutting; and Simon Lake received Patent No. 581,213 for "New and Useful Improvements in SubmarineVessels". Henry Ford received Patent No. 610,040 for "New and Useful Improvements in Carburetors" and later on in 1901, received Patent No. 686,046 for "New and Useful Improvements in Motor Carriages". It is worthy to note that Henry Ford was granted a total of 161 United States Patents, an impressive number by any standard, but hardly as impressive as the 1,093 patents which were issued to Thomas A. Edison.

Patents from 1900

Though it is difficult to spotlight which of the patents granted from 1900 to the present were most significant to the growth of our industry, certainly the "Flying Machine" of Orville and Wilbur Wright; "The Production of Bakelite" of Leo H. Baekeland; "The Neutronic Reacter" of Enrico Fermi and Leo Szilard; "The Rocket Motor" of Robert H. Goddard; "The Xerox Machine" of Chester Carlson and "The Manufacture of Nylon" by W. H. Carothers were among the most significant.

On May 23, 1930, a law was passed giving a person the right to obtain a patent on a plant if he invented or discovered and asexually reproduced any distinct and new variety of plant other than a tuberpropagated plant. This law aided agriculture and stimulated invention in new types of plants and the Act was approved by such notable men as Luther Burbank and Thomas Edion. In 1931, the first plant patent issued to Henry F. Bosenberg, of New Brunswick, New Jersey, for a climbing ever-blooming rose.

Patent laws over the years have been modified and changed to some extent, but the basic principles upon which the laws have been enacted are still the same and certainly Thomas Jefferson, the first supervisor of the American patent system, would be justly proud if he were to return today and see the fruits of his initial endeavors. The aspirations of the great American inventors, some of whom are above noted, should be an inspiration for all of us.

L. PASTEUR.
Manufacture of Beer and Yeast.
No. 141,072.   Patented July 22, 1873.

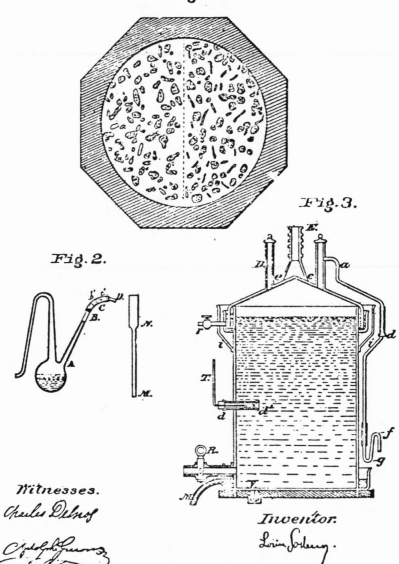

Fig.1.

Fig.2.

Fig.3.

Witnesses.
Charles Delnof

Inventor.
Louis Pasteur.

15 / PROTECTING THE INVENTION

Patents

A patent is, in effect, a contract agreement with the government whereby the inventor is granted a patent in return for public release of his invention in the form of a printed document. The patent is a limited monopoly for a period of years depending upon the country in which the patent is granted. It does not give the inventor the right to manufacture, use or sell but rather it gives the patentee the right to prevent others from manufacturing, using or selling. An inventor has the right to manufacture, use and sell only, if he is not infringing a live dominating patent. All patented inventions fall into the public domain upon termination of the life of the patent. This means that once the patent has expired, anyone can make the invention without fear of infringement.

Historically, the first grant for a creative idea affording protection to the creator was awarded by the Greeks several centuries B. C. in the culinary art. During the Middle Ages, the glassmakers of Venice received patents on their glassmaking processes and Galileo received a patent for an irrigating system in 1594.

For the most part, the present systems for securing patents are based on the British system which was established early in the 17th century. In the United States, Thomas Jefferson was instrumental in establishing a patent system for which provision was made in Article I, Section 8 of the United States Constitution, as mentioned in Chapter 14.

The first numbered British patent was granted in 1617 and the first United States patent was granted in 1790. Many other countries followed in succession to grant patents based on a system providing for an examination before issuing a patent to the inventor. This examination system is used today in many countries of the world and has resulted in the classification system of invention discussed in Chapter 3.

The total number of patents granted by all of the countries in the world runs into the millions. These patents enable the inventor to protect his creation against unlawful use and are treated as property in the same manner as real estate and personalty in that the patents may be licensed, sold or assigned for periods of time.

Patents have been the foundation stones of most of the industrial corporations in the world today. A patent gives to the individual a footing in our industrial complex if he should decide to establish his own business. A patent is on its face valid but the determination of validity may be questioned and a court can hold the patent invalid for many reasons such as fraud, finding that the invention has already been practiced earlier by someone else or finding that a printed publication or use or sale in public was made in advance of an established statutory date.

For the most part, patents comprise a description of the invention, drawings where applicable, and a claim or claims defining the invention. The scope and breadth of the claims determine whether the patent is a broad one covering a wide range or a narrow one covering only very specific details. A broad patent should include claims which vary in scope from the very broad to the specific.

Because the patent system and the laws which govern the granting of patents are complex, an inventor who is unskilled in the system and the laws should consult with a patent attorney cr agent. The United States Patent Office will furnish a list of attorneys and agents qualified to practice before it.

The disclosure

An inventor should keep careful and detailed notes of his research. He should keep them in a bound notebook in which the pages are numbered consecutively. He should date each of his entries and have them witnessed and dated by someone who understands the subject matter of his research. This information establishes a date of conception which may be very important in determining who was the first inventor when two or more inventors come up with the same idea. The inventor should also keep a careful record of the date on which he first reduced the idea to practice by making a working model or the like. A witness to this occasion and a record of the date of reduction to practice should be entered in his notebook.

The United States Patent Office furnishes detailed information on patents in a booklet entitled "General Information Concerning Patents" which can be procured by writing to the Commissioner of Patents in Washington, D. C. Other countries have similar programs and can be contacted through their Director or Commissioner of Patents. Patent attorneys may also provide you with additional information and it is well to consult with them as to problems which involve patenting, establishing a date of conception or a date of reduction to practice. It is important that an inventor show diligence in pursuing legal rights at all times from conception to filing an application.

An inventor should consider having a search conducted in order to determine whether his idea is novel. Such a search can be made through a patent attorney or by the individual if he makes a visit to the Patent Office of the country in which he wishes protection. A search is an invaluable tool to the inventor and it enables him to better evaluate his invention with respect to other developments in his field.

Determining the practicality and likelihood of success

One of the most difficult things for an inventor, or for that matter a corporation, or a patent lawyer, is to determine the practical aspects of an invention and its likelihood of commercial success. In general it can be said that the majority of patents realize some monetary return to the inventor. If the inventor works for a corporation, he is usually the recipient of some type of monetary reward as most corporations have provisions for inventors. If he is an individual, his chances of selling, licensing or producing his invention and making it successful are excellent. It is estimated that most patented developments are commercially successful either because they are actually manufactured or because they serve as a defense mechanism for a corporation to prevent others from manufacturing a particular device. Most corporations have defensive patents which protect a line of goods. The manufacturer might make one particular item but have patent protection on a number of related models which they do not manufacture. These patents prevent others from making competitive products in this specific field.

Most individual inventors become successful because they have a good marketing approach. An inventor may have a needed invention but without proper marketing, he may be unable to interest people in his invention. Therefore, an inventor should give serious thought and consideration to promoting his invention by someone skilled in this area. Success will depend a great deal on the marketing ability of the inventor or his marketing team. The old adage that a good marketing man can sell refrigerators to Eskimos holds true to inventions. Many items may appear to have little or no value, but when put in the hands of a good marketing man, they can be made successful commercial items.

Trade secrets and shop rights

An inventor may well consider the aspect of trade secrets. If he can manufacture the development without someone learning of the technique or secret behind its manufacture he might be well advised to keep the item as a trade secret and not have it patented. Some items can be kept secret for many years, but in general the inventor runs the risk of having it discovered by someone else and losing his patent rights.

As to shop rights, an inventor may have relinquished his patent rights to a company that he works for, since they have certain rights to inventions developed within the corporation. For questions involving shop rights and trade secrets it is well for the inventor to consult with a patent lawyer.

Copyrights

Copyright protects the writings of an author against copying. Literary, dramatic, musical and artistic works are included within the protection of the copyright law, which in some instances also confers performing and recording rights. The copyright goes to the form of expression rather than to the subject matter of the writing. A description of a machine could be copyrighted as a writing, but this would only prevent others from copying the description—it would not prevent others from writing a description of their own or from making and using the machine. Copyrights are registered in the Copyright Office in the Library of Congress and the Patent Office has nothing whatever to do with copyrights. Information concerning copyrights may be obtained by addressing: Register of Copyright, Library of Congress, Washington, D. C. 20540.

Trademarks

A trademark relates to any word, name, symbol or device which is used in trade with goods to indicate the source or origin of the goods and to distinguish them from the goods of others. Trademark rights may be used to prevent others from using a confusingly similar mark, but not to prevent others from making the same goods or from selling them under a non-confusing mark. Similar rights may be acquired in marks used in the sale or advertising of services (service marks). Trademarks and service marks which are used in interstate or foreign commerce may be registered in the Patent Office. The procedure relating to the registration of trademarks and some general information concerning trademarks is given in a pamphlet called General Information Concerning Trademarks, which may be obtained from The Commissioner of Trademarks, Washington, D.C. 20402, on request.

What can be patented

The first patent law was enacted in 1790. The law now in effect is a general revision which was enacted July 19, 1952, and which came into effect January 1, 1953. This law is reprinted in a pamphlet entitled "Patent Laws", which is sold by the Superintendent of Documents, U. S. Government Printing Office, Washington, D. C. 20402.

The patent law specifies the subject matter for which a patent may be obtained and the conditions for patentability.

The law establishes the Patent Office for administering the law relating to the granting of patents, and contains various other provisions relating to patents.

The patent law specifies the general field of subject matter that can be patented, and the conditions under which a patent may be obtained.

In the language of the statute, any person who "invents or discovers any new and useful process, machine, manufacture, or composition of matter, or any new and useful improvements thereof, may obtain a patent," subject to the conditions and requirements of the law. By the word "process" is meant a process or method, and new processes, primarily industrial or technical processes, may be patented. The term "machine" used in the statute needs no explanation. The term "manufacture" refers to articles which are made, and includes all manufactured articles. The term "composition of matter" relates to chemical compositions and may include mixtures of ingredients as well as new chemical compounds. These classes of subject matter taken together include practically everything which is made by man and the processes for making them.

The Atomic Energy Act of 1954 excludes the patenting of inventions useful solely in the utilization of special nuclear material or atomic energy for atomic weapons.

The statute specifies that the subject matter must be "useful." The term "useful" in this connection refers to the condition that the subject matter has a useful purpose and also includes operativeness, that is, a machine which will not operate to perform the intended purpose would not be called useful. Alleged inventions of perpetual motion machines are refused patents.

Interpretations of the statute by the courts have defined the limits of the field of subject matter which can be patented, thus it has been held that methods of doing business and printed matter cannot be patented. In the case of mixtures of ingredients, such as medicines, a patent cannot be granted unless there is more to the mixture than the effect of its components. (So-called patent medicines are ordinarily not patented; the phrase "patent medicine" in this connection does not have the meaning that the medicine is patented.) It is often said that a patent cannot be obtained upon a mere idea or suggestion. The patent is granted upon the new machine, manufacture, etc., as has been said, and not upon the idea or suggestion of the new machine. As will be stated later, a complete description of the actual machine or other subject matter sought to be patented is required.

In order for an invention to be patentable it must be new as defined in the statute. The statute provides that an invention cannot be patented if—

"(a) The invention was known or used by others in this country, or patented or described in a printed publication in this or a foreign country, before the invention thereof by the applicant for patent, or

"(b) The invention was patented or described in a printed publication in this or a foreign country or in public use or on sale in this country more than one year prior to the date of the application for patent in the United States. . . ."

If the invention has been described in a printed publication anywhere in the world, or if it has been in public use or on sale in this country before the date that the applicant made his invention, a patent cannot be obtained. If the invention has been described in a printed publication anywhere, or has been in public use or on sale in this country more than one year before the date on which an application for patent is filed in this country, a valid patent cannot be obtained. In this connection it is immaterial when the invention was made, or whether the printed publication or public use was by the inventor himself or by someone else. If the inventor describes the invention in a printed publication or uses the invention publicly, or places it on sale, he must apply for a patent before one year has gone by, otherwise any right to a patent will be lost.

Even if the subject matter sought to be patented is not exactly shown by the prior art, and involves one or more differences over the most nearly similar thing already known, a patent may still be refused if the differences would be obvious. The subject matter sought to be patented must be sufficiently different from what has been used or described before so that it may be said to amount to invention over the prior art. Small advances that would be obvious to a person having ordinary skill in the art are not considered inventions capable of being patented. For example, the substitution of one material for another, or changes in size, are ordinarily not patentable.

The United States Patent Office

The chief functions of the Patent Office are to administer

the patent laws as they relate to the granting of letters patent of inventions, and to perform other duties relating to patents. It examines applications for patents to ascertain if the applicants are entitled to patents under the law, and grants the patents when they are so entitled; it publishes issued patents and various publications concerning patents and patent laws, records assignments of patents, maintains a search room for the use of the public to examine issued patents and records, supplies copies of records and other papers, and the like. Analogous and similar functions are performed with respect to the registration of trademarks. The Patent Office has no jurisdiction over questions of infringement and the enforcement of patents, nor over matters relating to the promotion or utilization of patents or inventions.

The head of the Office is the Commissioner of Patents and his staff includes several assistant commissioners of patents and other officials. As head of the Office, the Commissioner superintends or performs all duties respecting the granting and issuing of patents and the registration of trademarks; exercises general supervision over the entire work of the Patent Office; prescribes the rules, subject to the approval of the Secretary of Commerce, for the conduct of proceedings in the Patent Office and for recognition of attorneys and agents; decides various questions brought before him by petition as prescribed by the rules, and performs other duties necessary and required for the administration of the Patent Office and the performance of its functions.

The examination of applications for patents is the largest and most important function of the Patent Office. The work is divided among a number of examining groups, each group having jurisdiction over certain assigned fields of invention. Each group is headed by a group director and staffed by a number of examiners. The examiners perform the work of examining applications for patents and determine whether patents can be granted. An appeal can be taken to the Board of Appeals from their decisions refusing patents and a review by the Commissioner of Patents may be had on other matters by petition. The examiners also determine when an interference exists between pending applications, or a pending application and a patent, institute interference proceedings in such cases and hear and decide certain preliminary questions raised by contestants.

Applications for patents are not open to the public, and

no information concerning them is released except on written authority of the applicant, his assignee, or his attorney, or when necessary to the conduct of the business of the Office. Patents and related records, including records of any decisions, the records of assignments other than those relating to assignments of patent application, books, and other records and papers in the Office are open to the public. They may be inspected in the Patent Office Search Room or copies may be ordered.

A Search Room is provided where the public may search and examine United States patents granted since 1836. Patents are arranged according to the Patent Office classification system of over 500 subject classes and 75,000 subclasses. By searching in these classified patents, it is possible to determine, before actually filing an application, whether an invention has been anticipated by a United States patent, and it is also possible to obtain the information contained in patents relating to any field of endeavor. The Search Room contains a set of United States patents arranged in numerical order and a complete set of the Official Gazette.

A Record Room also is maintained where the public may inspect the records and files of issued patents and other open records.

Since a patent is not always granted when an application is filed, many inventors attempt to make their own investigation before applying for a patent. This may be done in the Search Room of the Patent Office, and to a limited extent in some public libraries. Patent attorneys or agents may be employed to make a so-called preliminary search through the prior United States patents to discover if the particular device or one similar to it has been shown in some prior patent. This search is not always as complete as that made by the Patent Office during the examination of an application, but only serves, as its name indicates, a preliminary purpose. For this reason, the Patent Office examiner may, and often does, reject claims in an application on the basis of prior patents or publications not found in the preliminary search.

Who may apply for a patent

According to the statute, only the inventor may apply for a patent, with certain exceptions. If a person who is not the inventor should apply for a patent, the patent, if it were obtained, would be void. The person applying in such a case who falsely states that he is the inventor would also be subject to criminal penalties. If the inventor is dead, the application may be made by his legal representatives, that is, the administrator or executor of his estate, in his place. If the inventor is insane, the application for patent may be made by his guardian, in his place. If an inventor refuses to apply for a patent or cannot be found, a joint inventor or a person having a proprietary interest in the invention may apply on behalf of the missing inventor.

If two or more persons make an invention jointly, they apply for a patent as joint inventors. A person who makes a financial contribution is not a joint inventor and cannot be joined in the application as an inventor. It is possible to correct an innocent mistake in omitting a joint inventor or in erroneously joining a person as inventor.

Officers and employees of the Patent Office are prohibited by law from applying for a patent or acquiring, directly or indirectly, except by inheritance or request, any patent or any right or interest in any patent.

Models, exhibits, specimens

Models were once required in all cases admitting of a model, as a part of the application, and these models became a part of the record of the patent. Such models are no longer generally required since the description of the invention in the specification, and the drawings, must be sufficiently full and complete, and capable of being understood, to disclose the invention without the aid of a model. A model will not be admitted unless specifically called for.

A model, working model, or other physical exhibit, may be required by the Office if deemed necessary for any purpose on examination of the application. This is not done very often. A working model will be called for in the case of applications for patent for alleged perpetual motion devices.

When the invention relates to a compostion of matter, the applicant may be required to furnish specimens of the composition, or of its ingredients or intermediates, for inspection or experiment.

Interferences

Occasionally two or more applications are filed by different inventors claiming substantially the same patentable invention. The patent can only be granted to one of them, and a proceeding known as an "interference" is instituted by the Patent Office to determine who is the first inventor and entitled to the patent. About 1 percent of the applications filed become involved in an interference proceeding. Interference proceedings may also be instituted between an application and a patent already issued, provided the patent has not been issued for more than 1 year prior to the filing of the conflicting application, and provided that the conflicting application is not barred from being patentable for some other reason.

Each party to such a proceeding must submit evidence of facts providing when he made the invention. In view of the necessity of proving the various facts and circumstances concerning the making of the invention during an interference, inventors must be able to produce evidence to do this. If no evidence is submitted a party is restricted to the date of filing his application as his earliest date. The priority question is determined by a board of three interference examiners on the evidence submitted. From the decision of the Board of Patent Interferences, the losing party may appeal to the Court of Customs and Patent Appeals or file a civil action against the winning party in the appropriate United States district court.

The terms "conception of the invention" and "reduction to practice" are encountered in connection with priority questions. Conception of the invention refers to the completion of the devising of the means for accomplishing the result. Reduction to practice refers to the actual construction of the invention in physical form; in the case of a machine it includes the actual building of the machine, in the case of an article or composition it includes the actual making of the article or composition, in the case of a process it includes the actual carrying out of the steps of the process; and actual operation, demonstration, or testing for the intended use is also usually necessary. The filing of a regular application for patent completely disclosing the invention is treated as equivalent to reduction to practice. The inventor who proves to be the first to conceive the invention and the first to reduce it to practice will be held to be the prior inventor, but more complicated situations cannot be stated this simply.

Nature of patent and patent rights

The patent is issued in the name of the United States under the seal of the Patent Office, and is either signed by the Commissioner of Patents or has his name written thereon and attested by an official of the Patent Office. The patent contains a grant to the patentee and a printed copy of the specification and drawing is annexed to the patent and forms a part of it. The grant to the patentee is of "the right to exclude others from making, using or selling the invention throughout the United States" for the term of 17 years. The United States in this phrase includes Territories and possessions.

The exact nature of the right conferred must be carefully distinguished, and the key is in the words "right to exclude" in the phrase just quoted. The patent does not grant the right to make, use, or sell the invention but only grants the exclusive nature of the right. Any person is ordinarily free to make, use, or sell anything he pleases, and a grant from the Government is not necessary. The patent only grants the right to exclude others from making, using, or selling the invention. Since the patent does not grant the right to make, use, or sell the invention, the patentee's own right to do so is dependent upon the rights of others and whatever general laws might be applicable. A patentee, merely because he has received a patent for an invention, is not thereby authorized to make, use or sell the invention if doing so would violate any law. An inventor of a new automobile would not be entitled to use his new automobile in violation of the laws of a State requiring a license, because he has obtained a patent, nor may a patentee sell an article the sale of which may be forbidden by a law, merely because he has obtained a patent. Neither may a patentee make, use or sell his own invention if doing so would infringe the prior rights of others. A patentee may not violate the Federal antitrust laws, such as by resale price agreements or entering into combinations in restraint of trade, or the pure food and drug laws, by virtue of his having a patent.

Ordinarily there is nothing which prohibits a patentee from making, using, or selling his own invention, unless he thereby infringes another's patent which is still in force.

Since the essence of the right granted by a patent is the right to exclude others from commercial exploitation of the invention, the patentee is the only one who may make, use, or sell his invention. Others may not do so without authorization from the patentee. The patentee may manufacture and sell the invention himself or he may license, that is, give authorization to others to do so.

After the patent has expired anyone may make, use, or sell the invention without permission of the patentee, provided that matter covered by other unexpired patents is not used. The term may not be extended except by special act of Congress.

Infringement of patents

Infringement of a patent consists in the unauthorized making, using, or selling of the patented invention within the territory of the United States, during the term of the patent. If a patent is infringed, the patentee may sue for relief in the appropriate Federal court. He may ask the court for an injunction to prevent the continuation of the infringement, and he may also ask the court for an award of damages because of the infringement. In such an infringement suit, the defendant may raise the question of the validity of the patent, which is then decided by the court. The defendant may also aver that what he is doing, does not constitute infringement. Infringement is determined primarily by the language of the claims of the patent and, if what the defendant is making does not fall within the language of any of the claims of the patent, he does not infringe.

Suits for infringement of patents follow the rules of procedure of the Federal courts. From the decision of the district court, there is an appeal to the appropriate Federal court of appeals. The Supreme Court may thereafter take a case by writ of certiorari. If the United States Government infringes a patent, the patentee has a remedy for damages in the Court of Claims of the United States. The Government may use any patented invention without permission of the patentee, but the patentee is entitled to obtain compensation for the use by or for the Government.

If the patentee notifies anyone that he is infringing his patent or threatens suit, the one charged with infringement may himself start the suit in a Federal court and get a judgment on the matter.

The Patent Office has no jurisdiction over questions relating to infringement of patents. In examining applications for patent no determination is made as to whether the invention sought to be patented infringes any prior patent. An improvement invention may be patentable, but it might infringe a prior unexpired patent for the invention improved upon, if there is one.

The above is but a brief review of the rights and approaches available to an inventor. Many books have been written on this subject and detailed information can be obtained at most libraries.

In conclusion, it is suggested that an inventor check with The Patent Office as to general procedures and request that The Patent Office furnish him with whatever information is available relative to methods for protecting his inventions. It is also suggested that he consult with a patent lawyer in order to insure that he has not relinquished any rights to his inventions. An inventor should act promptly in this regard and not delay since he might, because of long delays, jeopardize his inventions.

ADDENDUM

About the Author

Mr. Shlesinger, was born in Rochester, New York on May 3, 1924. His father is an inventor and patent lawyer.

At an early age he showed a deep interest in science and while in high school devised a submarine with hydrafoils. In the summer of 1942, he worked for Will Corporation in Rochester, a supplier of laboratory equipment and chemicals. With the money earned at Will Corporation, and his father's assistance, he had built and equipped a substantial laboratory in his home.

His formal education began in 1942 at Holy Cross College in Worcester, Massachusetts. In early 1943, he enlisted in the Air Force and after attending numerous training schools including Furman University in Greenville, South Carolina, he was graduated a Navigator in 1944 and subsequently was sent to England where he flew in combat with the 8th Air Force in B-17s. While in the Air Force, he devised a system for determining the exact position of a bomber formation by using sound transmitting devices in anti-aircraft shells. The system was applicable to bombers of that period and was incapable of being jammed, but with the advent of jet aircraft, the system was never fully developed. After the war, he maintained his commission with the Air Force Reserve and served as Adjutant, Personnel Officer and Training Officer prior to his retirement in 1969 with the rank of Major. During his career with the Air

Force Reserve, he assisted in the development and augmentation of numerous training programs and submitted to the Army for consideration, an anti-personnel weapon capable of protecting a tank from being disabled by a person who had boarded the tank or gotten close enough thereto as to be under the range of depression of its guns.

After the end of World War II, Mr. Shlesinger returned to Holy Cross, received his degree and proceeded to graduate school at the University of Rochester. During his summers, he worked at Eastman Kodak Company in Rochester, New York, and in 1949 went to work for Kodak on a full-time basis. While at Kodak, he devised a method for recovering waste solvent for which he received an award.

In 1949, he married Rita Jane Belmont and shortly thereafter left Kodak to consider other job opportunities. He worked in his father's Rochester law office part-time until 1950 when he departed to attend law school in Washington, D. C., the area where he and his wife and their six children are presently living. Upon arrival in Washington, he went to work for Edwin B. Gary, a patent lawyer who died in 1951. That same year he was placed in charge of a Washington office opened by his father. This office has grown from a one man operation until today it is a large firm headed by Mr. Shlesinger, handling both national and international patent, trademark and copyright matters.

Mr. Shlesinger received his Juris Doctor degree from George Washington University and is a member of the bars of the U. S. Supreme Court, the U. S. Court of Customs and Patent Appeals, the U. S. Court of Claims, the U. S. Circuit Court of Appeals for the District of Columbia, the U. S. District Courts for the District of Columbia and the Eastern District of Virginia, the Supreme Court of Appeals of the Commonwealth of Virginia, and is registered to practice in the U. S. and Canadian Patent Offices.

While attending law school at night, and operating the Washington office during the day, Mr. Shlesinger continued with his development work. In 1954, he was granted his first patent on an adjustable stapling machine which was subsequently assigned to Bostich.

His early inventions included many items, such as a bailer for boats, a boxing ring and numerous developments in the area of stapling machines, some of which were patented. One early development was a top filling fountain pen for which he received a patent which was subsequently assigned to Scripto, Inc. In 1957, when the first Sputnik was put up by the Russians, he devised a launching method for rockets, for which he received a patent. His system was identical to a system subsequently used by Russians in the Ural Mountains.

In the late 50's, he became interested in electrical switching devices and was hired for a number of years as a consultant to AMP, Inc. of Harrisburg, Pennsylvania. His development work included the areas of programming switches, electrical connectors, fluid couplings, explosive couplings, and medical appliances and devices including a joint invention with Dr. Carlos E. Odiaga, a member of the American College of Surgeons, for use in connecting intestines, portions of which had been removed.

In his career as an inventor, he also collaborated with George A. Arkwright, one of his law partners and a mechanical engineer, to develop several programming systems and an anti-theft device to prevent removal of soft-drink bottles from racks.

In more recent years, the bulk of Mr. Shlesinger's developments have been in the field of reed switch devices and for several years he has collaborated with Charlie D. Mariner, also an inventor, in the development of the world's smallest reed switch.

Mr. Shlesinger's inventions can be found in many areas of present day use including electronic organs, computers, telephone switch gear and space equipment.

In the late 60's, Mr. Shlesinger conceived the idea of teaching a course on invention. After much study and research, he concluded that there was no one course or book or series of courses or books which taught invention by a simple step-by-step process. He decided that such a system was feasible and after much deliberation, put together a course outline. He was convinced that the outline was a workable one, but it had to be tested to prove practical. He discussed the matter with the Dean of the Engineering School of George Washington University and the course was worked into the Continuing Education Program of the Engineering School.

The original course outline went through many revisions and Mr. Shlesinger finally decided that he needed to test the step-by-step approach to invention on those who had no formal education. He discussed the matter with Dr. Stuart Adams of the Department of Corrections for the District of Columbia, and with his assistance, was able to present the course to the prisoners at Lorton, the District of Columbia correctional facility, located in Virginia. The course proved to be effective when presented even to men who had no formal education. Though some of his students were illiterate, by using the step-by-step method approach to invention, they were able to devise original and novel inventions.

In addition to teaching the prisoners at Lorton, he also presented the course at the Woodbridge Prison Facility of the State of Virginia. The results of these courses proved that invention could be taught to even those who had little or no education and each new group of students contributed towards the revision of the basic course outline. Additionally, seminars were conducted which materially added to the development of the course.

In 1971, Mr. Shlesinger began to put his course into a book form. This book was completed in December 1973. To prove its versatility, Mr. Shlesinger used portions of it for the first time in the fall of 1973 when he taught his course to a group of high school juniors and seniors at Bishop Ireton High School in Alexandria, Virginia. The results proved conclusively that the book is a basic tool for all those who wish to invent. Not one of the students thought himself capable of inventing when he started the course. The students were given simple tools such as the hammer, screwdriver, saw, wrench, and scissors as work projects. All of the students were able to invent new devices in their project area. Of the 30 odd inventions which they made, some unbeknown to them had been invented, but many were unique and in at least three instances, the inventions which they made deserved special merit as being highly original, novel and noteworthy contributions to the field of science.

In more recent years, Mr. Shlesinger has set up invention programs for students and teachers of elementary schools in Virginia, New Jersey, and the District of Columbia. His work has included children from kindergarten through eighth grades, gifted and talented children, and children in juvenile delinquent facilities.

Mr. Shlesinger has commented relative to this book that he wishes he had had such a tool when he was 17 or 18. Although he has invented over 1,000 devices and has over 100 United States Patents and numerous foreign patents, it required many years of research to realize that inventing could be systematized and made easy. He had to find this out the hard way.

INDEX